Frontiers in Nuclear and Particle Physics

(*Volume 1*)

Multi-electronic Processes in Collisions Involving Charged Particles and Photons with Atoms and Molecules

Edited by

Antônio Carlos Fontes dos Santos

Instituto de Fisica, Universidade Federal do Rio de Janeiro, Caixa Postal 68525, 21941-972, Rio de Janeiro, RJ, Brazil

Frontiers in Nuclear and Particle Physics

Volume # 1

A Multi-electronic Processes in Collisions Involving Charged Particles and Photons with Atoms and Molecules

Editor: Antônio Carlos Fontes dos Santos

ISSN (Online): 2589-756X

ISSN (Print): 2589-7551

ISBN (Online): 978-1-68108-613-2

ISBN (Print): 978-1-68108-614-9

General:

1. Any dispute or claim arising out of or in connection with this License Agreement or the Work (including non-contractual disputes or claims) will be governed by and construed in accordance with the laws of the U.A.E. as applied in the Emirate of Dubai. Each party agrees that the courts of the Emirate of Dubai shall have exclusive jurisdiction to settle any dispute or claim arising out of or in connection with this License Agreement or the Work (including non-contractual disputes or claims).
2. Your rights under this License Agreement will automatically terminate without notice and without the need for a court order if at any point you breach any terms of this License Agreement. In no event will any delay or failure by Bentham Science Publishers in enforcing your compliance with this License Agreement constitute a waiver of any of its rights.
3. You acknowledge that you have read this License Agreement, and agree to be bound by its terms and conditions. To the extent that any other terms and conditions presented on any website of Bentham Science Publishers conflict with, or are inconsistent with, the terms and conditions set out in this License Agreement, you acknowledge that the terms and conditions set out in this License Agreement shall prevail.

Bentham Science Publishers Ltd.
Executive Suite Y - 2
PO Box 7917, Saif Zone
Sharjah, U.A.E.
Email: subscriptions@benthamscience.org

BENTHAM SCIENCE

CONTENTS

FOREWORD

This e-book explores some of the most exciting frontiers of atomic, molecular and nuclear physics. Written and illustrated in a simplified manner, this e-book aims to provide insight to its readers how electrons play a major role in the atomic and nuclear collision processes. Indeed, this book contains a wealth of theoretical information and variety of fascinating experiments to unravel the fundamental mechanisms positron, negative, ion, photons, protons, and antiprotons which makes the readers visualize different types of collisions and the physics behind. Each of the nine chapters in this bool are written by prominent scientists from atomic and molecular physics including Prof. R. D. DuBois and A. C. F. Santos.

Narayan P. Appathurai
Scientist
Canadian Light Source Inc.

This eBook covers several essential aspects of multiple interaction processes between electrons, nucleus, atoms and molecules. It addresses exciting applications of these fundamental processes, and gives insights in the recent experimental and theoretical developments of leading research groups of these areas. It will provide the researchers of this field a valuable overview of this broad and rapidly evolving field. The various contributions, mostly from Latin-American research groups, show the dynamism of this community in the area of basic molecular processes. The forthcoming facility "Syrius Project" in Brazil will benefit from the expertise of this community and allow the realization of ground-breaking experiments. We hope that this eBook will promote interactions between the researchers of this area and provoke many passionate discussions.

Nelson V. de Castro Faria
Emeritus Professor of Physics,
Federal University of Rio de Janeiro

PREFACE

This eBook contains nine contributions, covering a diversity of subject areas in atomic, molecular and nuclear physics. Robert (Bob) DuBois surveys the ionization by antiparticles and particles. He compares fully to total differential ionization cross sections and finds where the differences are. His contribution concentrates on the progress and methods used in particles and antiparticles based studies of inelastic atomic interactions. This chapter shows how new experimental techniques are critical in advancing experimental studies from total or integral cross section measurements to highly differential investigations that are now being performed. The primary emphasis is on ionization of atoms and simple molecules by low-energy positrons and in showing similarities and differences between positron, electron and proton impact data. Experimental techniques associated with the generation, moderation, and transport of low-energy positron beams summarizing existing experimental studies are provided. Comments with respect to future studies and directions, plus how they might be achieved, are presented.

The chapter by Alcantara *et al.* deals with photoionization of an important atmospheric molecule, the dichloromethane. Chemical processes are of significant importance in determining the equilibrium distribution of ions in planetary ionospheres. Photoionization and impact ionization are responsible for the initial creation of the electron-ion pairs. Thus, the knowledge of the chemical processes taking place in the Earth`s atmosphere is necessary for the understanding of ionospheric structure and behavior. Thus, the chapter by Alcantara *et al.* presents the centre of mass kinetic energy release distribution spectra of ionic fragments, formed through dissociative single and double photoionization of CH_2Cl_2 at photon energies around the Cl 2p edge, which were extracted from the shape and width of the experimentally obtained time-of-flight distributions. The authors report that the kinetic energy distributions spectra exhibit either smooth profiles or structures. In the particular case of double ionization with the ejection of two charged fragments, the kinetic energy distributions present own characteristics compatible with the Coulombic fragmentation model. Intending to interpret the experimental data singlet and triplet states at Cl 2p edge of the CH_2Cl_2 molecule, corresponding to the resonant transitions, were calculated at multiconfigurational self-consistent field level and multireference configuration interaction. Minimum energy pathways for dissociation of the molecule were additionally calculated aiming to support the presence of the ultra-fast dissociation mechanism in the molecular break-up.

Paulo Limão-Vieira and colleagues revisit electron transfer processes yielding negative ion formation in gas-phase collisions of fast neutral potassium atoms (electron donor) and biomolecular target molecules (electron acceptor) in a crossed molecular-beam arrangement. The negative ions formed in the interaction region are time-of-flight (TOF) mass analysed as a function of the collision energy. Selective site and bond excision in the unimolecular decomposition of the transient negative show clear dependence on the collision energy.

The chapter by Santos and Almeida aims to the study of electron correlation in atoms, which is a significant but difficult study subject to its many-body interactions, nevertheless it may be the main process to describe the absolute electron impact cross section for the ionization of atoms and molecules. The outer-shell double photoionization of a multi-electron target is usually a weak process in comparison to single photoionization and it is determined completely by electron-electron interaction. The main scope of this chapter has been upon a review of the recently found relationship between shake-off probabilities and target atomic number and electron density. By comparing the saturation values of measured double-t--single photoionization ratios from the literature, a simple scaling law was found, which allows us to forecast the shake-off probabilities for several atomic elements up to xenon within a factor two. The electron shake-off probabilities accompanying valence shell photoionization have been scaled as a function of the target atomic number, and static polarizability.

Claudia Montanari analyzes the multiple ionization by the impact of singly charged projectiles: electrons, positrons, protons and antiprotons. This comparison allows to study the mass and charge effects on the ionization processes. To that end Dr. Montanari includes the theoretical description given by the continuum distorted wave eikonal initial state, and a detailed compilation of the experimental data available for particle (proton and electron) and antiparticle (antiproton and positron) multiple ionization of the heaviest rare gases. The multiple ionization results include the Auger-type post-collisional contribution, which enhances the number of emitted electrons. For heavy projectiles, this is important at high energies. For light projectiles such as electrons and positrons, the Auger-type processes dominate the highly-charged ion production in the whole energy range, even close to the energy threshold. For this reason, the theoretical calculation of multiple ionization values requires good description of the deep-shell contributions.

In a work involving researchers from USA, Canada and Brazil, the fragmentation of the tetrachloromethane molecule, following core-shell photoexcitation and photoionization in the neighborhood of the chlorine K-edge is presented by using time-of-flight mass spectroscopy and monochromatic synchrotron radiation. Branching ratios for ionic dissociation were derived for all detected ions, which are informative of the decay dynamics and photofragmentation patterns of the core-excited species. In addition, the absorption yield has been measured, with a new assignment of the spectral features. The structure that appears above the Cl 1s ionization potential in the photoionization spectrum, has been ascribed in terms of the existing connection with electron-CCl_4 scattering through experimental data and calculations for low-energy electron-molecule cross sections. In addition, the production of the doubly ionized Cl^{2+} as a function of the photon energy has been analyzed in the terms of a simple and appealing physical picture, the half-collision model.

Tőkési and Varga present a theoretical description of the spectra of electrons elastically scattered from various samples. The analysis is based on very large scale Monte Carlo simulations of the recoil and Doppler effects in reflection and transmission geometries. Besides the experimentally measurable energy distributions the simulations give many partial distributions separately, depending on the number of elastic scatterings (single, and multiple scatterings of different types). Furthermore, they present detailed analytical calculations for the main parameters of the single scattering, taking into account both the ideal scattering geometry, i.e. infinitesimally small angular range, and the effect of the real, finite angular range used in the measurements. The effect of the multiple scattering on intensity ratios, peak shifts and broadening, is shown. The authors show results for multicomponent and double layer samples. Our Monte Carlo simulations are compared with experimental data.

Ginette Jalbert et al present a contribution that relates molecular and atomic physics with the fundamental of quantum mechanics. This chapter reports the accomplishment of successful measurements in detecting two metastable atoms H(2s) metastable atoms arriving from the dissociation of the same hydrogen molecule induced by electron impact. Two detectors, placed close to the collision center, measure the neutral metastable H(2s) through a localized quenching process, which mixes the H(2s) state with the H(2p), leading to a Lyman-α detection. The data show the accomplishment of a coincidence measurement which proves for the first time the existence of the H(2s) + H(2s) dissociation channel. These results may stimulate theoretical computations regarding the production of H(2s)+ H(2s) coming from the doubly excited states, by electron impact, of the molecular hydrogen. In addition, an emerging pair of atoms with mean lifetime of the order of 0.1 s and with polarized angular momentum (polarized spin or polarized total angular momentum) may provide a new manner to obtain insight into the complex field of the molecular interactions, from the short-distance to the long-distance domain of interactions between moving atoms. Besides, the authors present a proposal to test the spin coherence of molecular dissociation processes. Clearly, the H(2s) + H(2s) dissociation channel has a great potential as a coherent twin-atom source.

Finally, most of the knowledge of the atomic nucleus was obtained from experimental data involving stable nuclei or nuclei in the vicinity of the stability line. Since the 1980s, several intermediate energy laboratories in the world started to produce nuclei out of the stability line, Rare Ion Beams. Many new interesting phenomena related to these nuclei have been discovered so far. Light nuclei far away from the stability line such as ^{6,8}He, ^{11}Be, ^{11}Li, ^{22}C, ^{24}O and others have been produced in laboratory. Some of these nuclei present a pronounced cluster structure formed by a core plus one or more loosely bound neutrons forming a kind of low density nuclear matter around the core (nuclear halo). Most of the research involving RIB was developed at intermediate energies, from 30 up to hundreds of MeV/nucleon, and more recently, some facilities are producing secondary beams to perform scattering experiments at energies around the Coulomb barrier. Heavy ion elastic scattering angular distributions at incident energies close to the Coulomb barrier, when plotted as a ratio to the Rutherford cross section, frequently exhibit a typical Fresnel type diffraction pattern, with oscillations in the forward angle region. This behaviour is a consequence of the interference between the Coulomb and nuclear scattering amplitudes. Due to the low binding energies of exotic projectiles, the coupling between the elastic channel and the breakup states of the projectile is very important and strongly affects the elastic angular distributions, with a damping of the Fresnel oscillations and the complete disappearance of the Fresnel peak in some cases. To describe the effect of the breakup of the projectile in the elastic scattering, new theoretical approaches have been developed. Viviane Morcele and colleagues present elastic scattering angular distributions at energies a little above the Coulomb barrier. The angular distributions have been analyzed by Continuum-DiscretizedCoupled-Channels calculations to take into account the effect of the ^{6}He breakup on the elastic scattering. Two different approaches were used to describe the structure of the projectile. One considering the ^{6}He as a three-body system consisting of an alpha particle and two neutrons which, in addition to the target, form a four-body problem. To compare, in a second approach, the projectile is described as a two-body cluster formed by an alpha particle plus a di-neutron. A new kind of effect due to the projectile breakup in the elastic scattering angular distributions is reported.

I would like to thank the authors for their contributions and cooperation in assembling this ebook.

Antônio Carlos Fontes dos Santos
Instituto de Fisica, Universidade Federal do Rio de Janeiro,
Caixa Postal 68525, 21941-972,
Rio de Janeiro, RJ,
Brazil

List of Contributors

A. Medina	Instituto de Física, UFBA, Salvador, BA 40210-340, Brazil
A.B. Rocha	Instituto de Química, Universidade Federal do Rio de Janeiro - 21941-909, Rio de Janeiro, RJ, Brazil
A.C.F. Santos	Instituto de Física, Universidade Federal do Rio de Janeiro - 21941-972, Rio de Janeiro, RJ, Brazil
A.H.A. Gomes	Instituto de Física Gleb Wataghin, Universidade Estadual de Campinas - 13083-859, Campinas, SP, Brazil
C.C. Montanari	Instituto de Astronomía y Física del Espacio, Consejo Nacional de Investigaciones Científicas y Técnicas, Universidad de Buenos Aires, casilla de correo 67, sucursal 28, C1428EGA, Buenos Aires, Argentina
C.R. de Carvalho	Instituto de Física, UFRJ, Cx. Postal 68528, Rio de Janeiro, RJ 21941-972, Brazil
D. Varga	Institute for Nuclear Research, Hungarian Academy of Sciences (MTA Atomki), H–4001 Debrecen, P.O. Box 51, Hungary, EU
D.P. Almeida	Departamento de Física, Universidade Federal de Santa Catarina, Florianópolis, 88040-900, Brazil
F. Ferreira da Silva	Laboratório de Colisões Atómicas e Moleculares, CEFITEC, Departamento de Física, Faculdade de Ciências e Tecnologia, Universidade Nova de Lisboa, 2829-516 Caparica, Portugal
F. Impens	Instituto de Física, UFRJ, Cx. Postal 68528, Rio de Janeiro, RJ 21941-972, Brazil
F. Zappa	Departamento de Física, UFJF, MG 36036-330, Brazil
G. García	Instituto de Fisica Fundamental, Consejo Superior de Investigaciones Científicas, Serrano 113-bis, 28006 Madrid, Spain
G. Jalbert	Instituto de Física, UFRJ, Cx. Postal 68528, Rio de Janeiro, RJ 21941-972, Brazil
G.G.B. Souza	Instituto de Quimica, Universidade Federal do Rio de Janeiro, Rio de Janeiro, RJ, 21949-900, Brazil
J. Robert	Lab. Aimée Cotton CNRS, Univ. Paris Sud-11/ENS Cachan - 91405 Orsay, France
K. Tőkési	Institute for Nuclear Research, Hungarian Academy of Sciences (MTA Atomki), H–4001 Debrecen, P.O. Box 51, Hungary, EU
K.F. Alcantara	Instituto Nacional de Tecnologia, 20081-312. Rio de Janeiro, RJ, Brazil
K.T. Leung	Department of Chemistry, University of Waterloo, N2L 3G1, Canada
L. Sigaud	Instituto de Física, Universidade Federal Fluminense - 24210-346, Niterói, R.J., Brazil
L.O. Santos	Instituto de Física, UFBA, Salvador, BA 40210-340, Brazil
M. Rodríguez-Gallardo	Departamento de Física Atómica, Molecular y Nuclear, Universidad de Sevilla, Apdo. 1065, E-41080, Sevilla, Spain

M.M. Sant'Anna	Instituto de Fisica, Universidade Federal do Rio de Janeiro,Caixa Postal 68525, 21941-972, Rio de Janeiro, RJ, Brazil
N.V. de Castro Faria	Instituto de Física, UFRJ, Cx. Postal 68528, Rio de Janeiro, RJ 21941-972, Brazil
P. Limão-Vieira	Laboratório de Colisões Atómicas e Moleculares, CEFITEC, Departamento de Física, Faculdade de Ciências e Tecnologia, Universidade Nova de Lisboa, 2829-516 Caparica, Portugal
R. Lichtenthäler	Instituto de Física da Universidade de São Paulo, Depto. de Física Nuclear, CEP. 05508-090, S\~ao Paulo, Brazil
R.D. DuBois	Missouri University of Science and Technology, Rolla, MO, USA
V. Morcelle	Departamento de Física, Universidade Federal Rural do Rio de Janeiro, CEP. 23890-000, Rio de Janeiro, Brazil
W. Wolff	Instituto de Física, Universidade Federal do Rio de Janeiro - 21941-972, Rio de Janeiro, RJ, Brazil
W.C. Stolte	Advanced Light Source, Lawrence Berkeley National Laboratory, Berkeley, CA, 94720, USA

Antimatter-Matter and Matter-Matter Atomic Interactions: Their Similarities and Differences

DuBois R.D.[*]

Missouri University of Science and Technology, Rolla, MO USA

Abstract: Total and differential cross sections for positron and electron impact on argon atoms are compared in order to show their similarities and differences. These comparisons provide information as to how antimatter-matter atomic interactions are like, or different from, matter-matter interactions that are normally encountered. Plus such comparisons provide information about how simply changing the direction of the coulomb field in atomic interactions influences the interaction probabilities and the kinematics. Data taken from the literature are used for these comparisons. The selected data are considered to be the most reliable available and representative of the many studies performed to date.

Keywords: Antimatter, Atomic collisions. Positron impact, Charge effects, Coulomb interaction, Differential cross sections, Elastic collisions, Electron impact, Electron correlation, Ionization, Inelastic collisions, Shake-off, TS-1.

I. INTRODUCTION

Many types of atomic interactions take place as charged atomic particles impact on or travel through gases, liquids and solids. These range from elastic scattering, where only the projectile momentum is altered, to highly inelastic processes, where a significant amount of the projectile kinetic energy and momentum can be transferred to one or more target electrons. Beginning over a century ago, many electron beam studies have been performed to investigate such processes. With the discovery of the positron, new possibilities became possible. These include the ability to enhance our understanding of matter-matter interactions since for positron impact certain interactions such as electron exchange are prohibited. Thus, comparing electron and positron impact data can show the importance of this process. Also, comparison of positron and electron impact data can provide

---

[*] **Corresponding author R.D. DuBois:** Missouri University of Science and Technology, Rolla, MO USA; Tel: +1 573 341 -4781; E-mail: dubois@mst.edu

Antônio Carlos Fontes dos Santos (Ed.)

information about how simply reversing the Coulomb forces alters various interaction probabilities and reaction kinematics. Such information can be used to test and improve existing theoretical models of matter-matter interactions. In addition, comparison of positron and electron induced processes provides insight into the similarities and differences of antimatter-matter and matter-matter interactions.

Here, we will compare positron and electron impact data for interactions with argon atoms using a range of experimental information currently available. The primary reason for choosing interactions with argon atoms is because many positron based studies with argon have been performed. Another reason is that any similarities and differences observed for argon should also be representative of interactions involving other multi-shell, medium size atoms, thus a somewhat broader picture can be obtained.

Fig. (**1**) shows a simple schematic picture of how positrons and electrons behave differently as they approach, pass by, and depart an atom. There are two major differences, both of which are associated with the sign of the projectile charge. One difference is a trajectory effect, where because the coulomb force between the projectile and partially screened nucleus is attractive/repulsive for electrons/positrons, the impact parameter will be smaller/larger for otherwise equivalent incoming particles. This means that the probability of interaction as well as the interaction kinematics will differ. Another trajectory effect occurs post-collision. Depending on the sign of the projectile charge, the scattered particle and the ejected electron will either be attracted to or repelled from each other.

The other major difference is a polarization effect where again because of the opposite direction of the coulomb forces, the target electron cloud is attracted to or pushed away from the interacting particle. Again, this will influence the interaction probabilities and to some extent the final kinematics. Please note that for antimatter-matter lepton ionizing interactions, *e.g.*, for positron impact, the outgoing particles are distinguishable. Thus the kinematics can be studied in detail. In contrast, for matter-matter ionizing interactions by leptons, *e.g.*, for electron impact, kinematic arguments must be used to "identify" the outgoing particle. Other differences, not shown in Fig. (**1**) but easy to visualize, include the presence or absence of certain channels such as electron exchange (present only for electron impact) or capture (present only for positron impact).

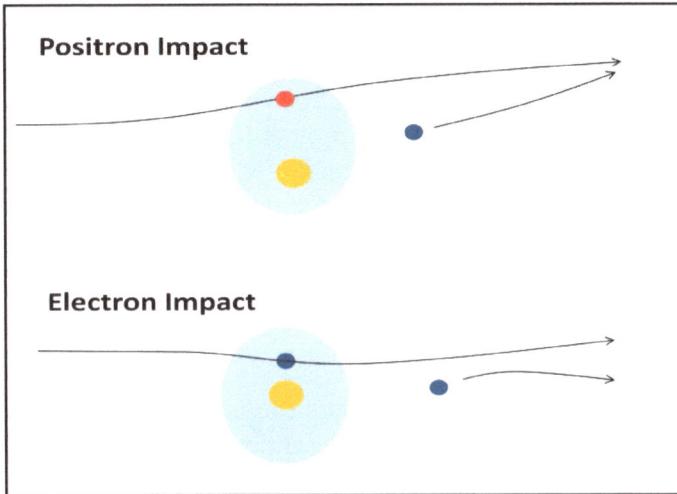

Fig. (1). Schematic showing target polarization and trajectory effects for positron and electron impact.

From a theoretical viewpoint, the attractive/repulsive trajectory effects are associated with scattering from the target core potential. This results in a repulsive force for positrons and an attractive force for electrons. This core scattering term must be combined with a polarization term which is attractive for positrons and repulsive for electrons. At low impact energies these terms tend to cancel for positron impact and add for electron impact whereas at sufficiently high energies, the polarization term becomes unimportant and only the static core interaction remains. These imply larger interaction cross sections for electron impact at low energies and identical cross sections at high energies.

Using a physical picture, the polarization and trajectory effects mean that the nuclear charge is more effectively screened for positron impact than it is for electron impact. Thus, one would expect a smaller elastic scattering cross section for positrons but a higher probability for inelastic interactions because more of the electron cloud is closer. However, the probability of electron exchange or capture must be added to obtain an overall picture. For electron impact electron exchange will further enhance both the elastic and inelastic cross sections. For positron impact, electron capture (Ps formation) should be significant at low energies because of the longer time that the positron and bound electrons are in close proximity of each other. Ps formation will greatly increase the overall inelastic cross section but will also rob available flux from the ionization channel. Thus, at lower energies physical arguments imply smaller elastic and ionization cross sections for positron impact. On the other hand, at significantly high impact energies the cross sections should be identical because the transverse forces leading to trajectory effects will be negligible compared to the incoming momenta and because the electron cloud does not have time to polarize.

Finally, theories using a perturbative expansion predict differences in cross sections due to the opposite signs of the projectile which influences certain terms. This is most evident for a 2nd Born expansion where the cross term scales as the third power of the product of the projectile and bound electron charges. This term is negative for positron impact and positive for electron impact. Thus, the double ionization cross section should be larger for electron impact as compared to positron impact, something that has been confirmed by many experimental studies. The reader is referred to Charlton *et al.* [1] for an example of this.

In the following sections, we will illustrate the influence of these features using total and differential data. As will be seen, under certain conditions there is little or no difference in how antimatter (positrons) and matter (electrons) interact with matter whereas for other conditions, large differences are seen. The data used in making these comparisons are taken from the literature and were selected to be representative or the most reliable data available. But the reader should keep in mind that the cross sections were measured using different techniques and in many cases were placed on absolute scales by normalizing to other measurements. Thus, agreement or disagreement within a 15-20% level should be viewed with caution as comparisons on the absolute level are subject to which set of normalization data was used. If a more detailed analysis than provided here is required, the reader should visit the references quoted.

II. TOTAL CROSS SECTION COMPARISONS

The first comparison we make is on the total (integral) cross section level. Cross sections for the various interaction channels associated with positron and electron impact on argon are shown in Fig. (**2**). For positron impact, the uppermost (black) curve is the total cross section for elastic plus inelastic interactions. This curve is a combination of the values recommended by Chiari and Zecca [2] and the measurements of Dababneh *et al.* [3]. Note that the early unpublished measurements of Coleman *et al.* [4] as quoted in Joachain *et al.* [5] are consistent with the later, more extensive, measurements of Dababneh *et al.* For inelastic interactions, the Ps production (electron capture) channel cross sections are shown by the blue curve. These data are also the recommended values of Chiari and Zecca [2]. The ionization (red) curve is from the combined measurements of Van Reeth *et al.* [6], Jacobsen *et al.* [7], Moxom *et al.* [8], Mori and Sueoka [9] and Kauppila *et al.* [10]. Lastly, the elastic cross section (magenta) curve is obtained *via* subtraction of the inelastic (ionization plus Ps production) cross sections from the total elastic plus inelastic cross section curves. For display purposes, the elastic cross section data for energies less than 7 eV have been shifted slightly downward.

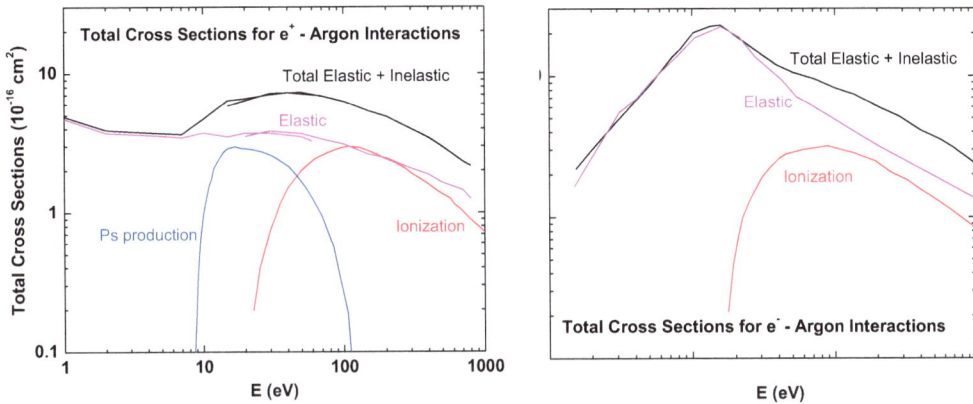

Fig. (2). Measured total cross sections for positron (left figure) and electron (right figure) -argon interactions. Positron impact: Total elastic plus inelastic (black curve), recommended values of Chiari and Zecca [2] and data of Dababneh *et al.* [3]; Ps production (blue curve), recommended values of Chiari and Zecca [2]; ionization (red curve), data of Van Reeth *et al.* [6], Jacobsen *et al.* [7], Moxom *et al.* [8], Mori and Sueoka [9], Kauppila *et al.* [10]. The elastic cross section curve (magenta) is obtained *via* subtraction with the data for energies less than 7 eV being shifted slightly downward for display purposes. Electron impact: Total elastic plus inelastic (black curve): suggested values from Gargioni Grosswendt [11]; total ionization (red curve): suggested values from Gargioni and Grosswendt [11], data of Straub *et al.* [12], Sorokin *et al.* [13], Wetzel *et al.* [14], Rapp and Englander-Golden [15] and Kauppila *et al.* [10]. Elastic (magenta curve): suggested values from Gargioni and Grosswendt [11], data of Gibson *et al.* [16], Iga *et al.* [16], Panajotović *et al.* [18], Srivastava *et al.* [19], Furst *et al.* [20] and DuBois and Rudd [21].

The right portion of the figure shows cross sections for electron impact. Here the total elastic plus inelastic cross sections (black curve) are the suggested values from Gargioni and Grosswendt [11]. The ionization cross sections are the combined values also suggested by Gargioni and Grosswendt [11] plus measurements of Straub *et al.* [12], Sorokin *et al.* [13], Wetzel *et al.* [14], Rapp and Englander-Golden [15] and Kauppila *et al.* [10]. The elastic scattering cross sections (magenta curve) are a combination of the suggested values by Gargioni and Grosswend [11] and the measurements of Gibson *et al.* [16], Iga *et al.* [17], Panajotović s [18], Srivastava s [19], Furst *et al.* [20] and DuBois and Rudd [21].

Before discussing comparisons between positron and electron impact, let us look at the overall characteristics of each. For both projectiles, elastic interactions dominate below 100 eV, more so in the case of electron impact than for positron impact. At high impact energies, the probabilities for elastic and inelastic interactions are comparable for positron impact but, for electron impact elastic collisions remain more probable. That the elastic and ionization cross sections are comparable for positron impact can be attributed to the reduced probability of elastic scattering due to the extra screening of the nuclear charge plus the increased probability of inelastic interactions due to the closer proximity of more of the electron cloud, as illustrated in Fig. (**1**). That elastic scattering tends to dominate for electron impact can be attributed to the same reasons. But for electron impact the screening is reduced, thus enhancing the elastic cross section. Ionization is also reduced because fewer electrons are near the projectile as it passes by. The other thing to note in Fig. (**2**) is that for positron impact the overall inelastic cross section, *i.e.*, the sum of the Ps formation and ionization channels, is significantly larger at low energies compared to the inelastic channel for electron impact. This is consistent with the picture where polarization causes more of the electron cloud to be near the positron as it passes by.

In Fig. (**3**) the same curves are plotted in order to compare the cross sections for positron (dashed curves) and electron impact (solid curves). The Ps formation curve is included in order to illustrate that this channel both decreases the threshold energy and significantly increases the cross section for inelastic interactions at lower energies for positron impact. Except at the very lowest energies shown, the cross sections for elastic scattering are significantly larger for electron impact than for positron impact.

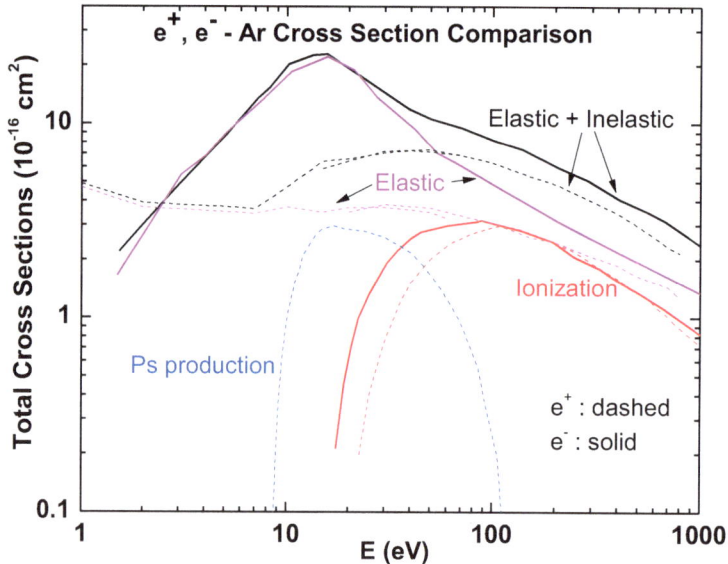

Fig. (3). Comparison of total cross sections for positron (dashed curves) and electron (solid curves) impact on argon. Data are the same as in Fig. (**2**).

This is consistent with the theoretical and physical arguments discussed in the introduction. Only at the very highest impact energies shown, do the elastic interaction probabilities seem to merge. Merging at higher energies occurs much sooner for ionizing interactions. At lower energies, namely below 100 eV, ionization of argon resulting in electron emission by electron impact is more probable than for positron impact. This is again in accordance with arguments discussed in the introduction. One should also note that the ionization cross sections deviate from each other in the same region where Ps formation is important.

This illustrates how the loss of flux to the capture channel aids in reducing the probability of direct emission of target electrons. Looking at the total elastic plus inelastic cross sections, the data imply that they still have not merged, even for impact energies two orders of magnitude larger than the ionization energy. Whether this indicates an overall normalization error, most likely for the positron impact data, or that the merging occurs at still higher energies is uncertain from data available at this time.

III. CROSS SECTION COMPARISONS FOR SINGLE AND DOUBLE ELECTRON REMOVAL

Let us now look a bit closer at the ionization channel. In Fig. (**4**) the cross sections for single (filled symbols) and double (open symbols) electron removal from argon by positrons (circles) and electrons (triangles and solid and dashed curves) are shown. Note that the double ionization cross sections include direct removal of two outer shell electrons plus removal of an inner shell electron followed by an Auger decay transition. The positron data are those of Jacobson *et al.* [7] and Bluhme *et al.* [22]. The Bluhme *et al.* data for impact energies less than 100 eV are not shown as their measurements include contributions from the Ps formation channel. The electron data are those of McCallion *et al.* [23] (solid and dashed curves) and Rejoub *et al.* [24] (filled and open triangles). These two data sets agree well for single ionization but for double ionization the McCallion *et al.* cross sections (the dashed curve) are about 15% larger than those reported by Rejoub *et al.* (the open triangles).

With regard to the differences in absolute cross sections, the general method of obtaining double ionization cross sections is to measure double to single cross section ratios and normalize these data using total ionization cross sections. Depending on the source used for the total cross sections, differences on the order of 10-15% in the absolute cross sections can result. However, from the many studies of double ionization (see ref 1, for example), the consensus is that the probability for removing a single target electron at high energies is the same for positron and electron impact while it is roughly twice more as likely that electron impact will result in double ionization. Thus, in Fig. (**4**) the dashed curve probably provides the best comparison with the positron double ionization data which are shown by the open circles. Also, with regard to the double ionization comparison, at energies about ~250 eV, L-shell ionization followed by an Auger decay contributes to the cross sections shown. Inner shell ionization cross sections have not been measured for positron impact so interpreting differences above this energy should be done with caution. Below approximately 100 eV, single electron removal by electron impact is more likely whereas there is little or no difference in the probability of double electron removal by positron or electron impact. Since single ionization dominates, the differences noted in Figs. (**3 and 4**) for single and total electron removal mimic each other.

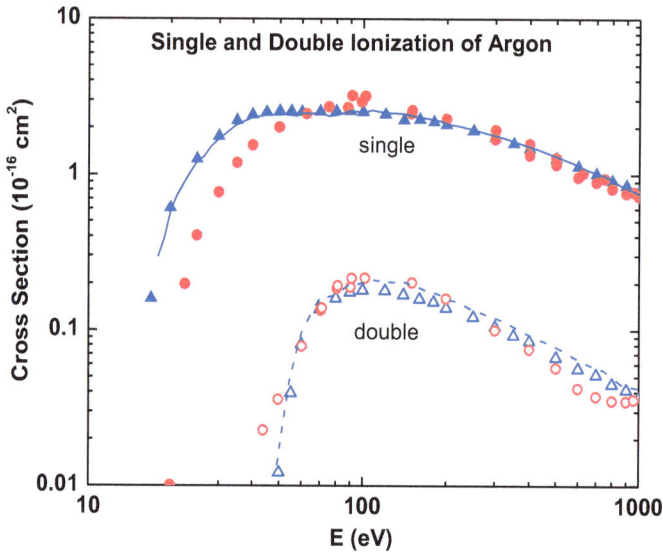

Fig. (4). Total cross sections for single and double ionization of argon by positrons (filled and open circles) and electrons (filled and open triangles plus solid and dashed curves). Positron data are from Bluhme *et al.* [22] and Jacobsen *et al.* [7]. Electron data are from McCallion *et al.* [23] (solid and dashed curves) and Rejoub *et al.* [24] (filled and open triangles).

IV. SINGLY AND DOUBLY DIFFERENTIAL CROSS SECTION COMPARISONS

Fig. (**3**) showed that the probabilities for ionizing argon by positron and electron impact are maximum and nearly identical around 100 eV. Also, at this energy and above the elastic and inelastic scattering probabilities have similar magnitudes. Because of this and since lepton beam experiments can be easily performed in the few hundred eV energy range, several types of differential studies have been performed in this region for positron impact. These data can be compared with the multitude of electron impact data available in order to gain greater insight into the kinematic similarities and differences associated with the sign of the projectile charge.

The first example of these kinematic features is shown in Fig. (**5**). Here differential cross sections for elastic scattering as a function of scattering angle are shown for 100 and 300 eV electron impact (the blue dashed curve and the filled stars and solid curve, respectively). The 100 eV data are those of DuBois and Rudd [21] while the 300 eV data are a combination of the measurements of Williams and Willis [25] and Jansen s [26]. These are compared to the positron impact data of Dou *et al.* [27] and Falke *et al.* [28]. Here, the Falke *et al.* data have been placed on an absolute scale by normalizing to the average value of integrated doubly differential cross sections for the sum of positron scattering plus electron emission reported by Kövér *et al.* [29] at a 30° observation angle.

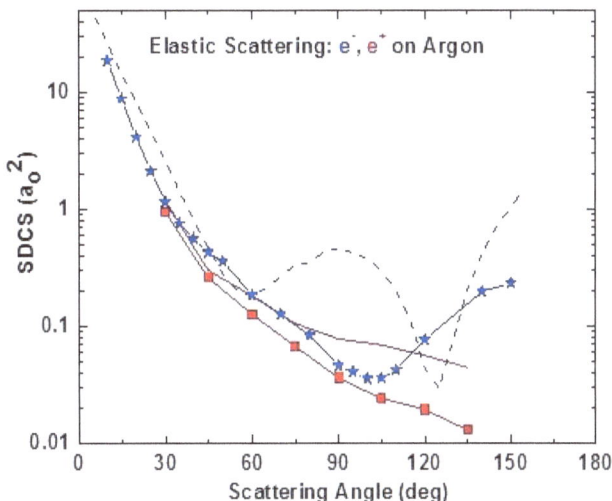

Fig. (5). Singly differential cross sections for elastic scattering from argon by positrons and electrons. 100 eV impact: dashed curve for electron impact, DuBois and Rudd [21]; solid curve for positron impact, Dou *et al.* [27]. 300 eV impact: filled squares and solid curve for positron impact, Falke *et al.* [28] normalized as described in the text; filled stars and solid curve, electron impact from Williams and Willis [25] and Jansen *et al.* [26].

As seen, the differential cross sections for forward scattering angles, *e.g.*, less than 60°, are very similar for positron and electron impact. But, at larger angles the probability that a positron scatters elastically decreases monotonically whereas electron elastic scattering has significant structure. In particular, there is a marked increase for electron scattering in the backward direction. In contrast, a similar behavior is totally absent when the projectile has the opposite charge.

Going back to Fig. (**1**) the differences in the backward direction can be attributed to an incoming electron being attracted toward the positively charged nucleus whereas a positron will be pushed further away. Thus, when the incoming particle (the electron) is closer there is a distinct possibility that it will be deflected completely around the nucleus and end up exiting in the backward direction. With regard to the differences in the total elastic cross sections, Fig. (**5**) shows that is primarily associated with scattering in the forward direction, which from the physical arguments presented in the introduction, is probably due to the differences in screening of the nuclear charge by the polarized electron cloud combined with trajectory deviations as the incoming lepton passes by the argon atom.

Turning our attention to the inelastic channel, Fig. (**6**) shows singly differential cross sections for an incoming lepton to be scattered or a bound argon electron to be ejected at a specific angle. Shown are electron impact data measured at 100 eV [30] compared to positron impact data measured at 90 and 120 eV [28]. The positron data have again been placed on an absolute scale by normalizing the integrated doubly differential cross sections provided in reference 29.

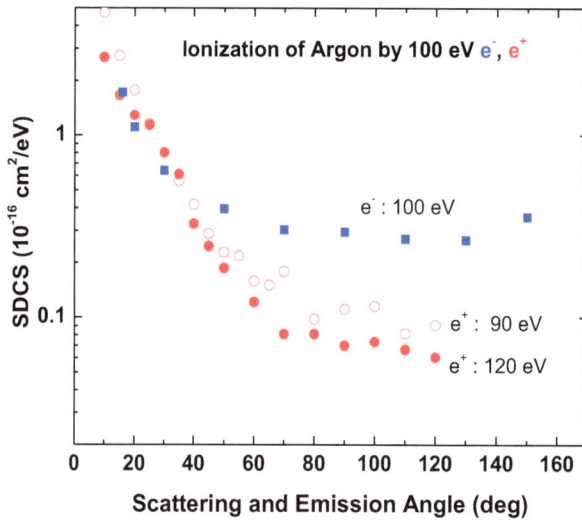

Fig. (6). Singly differential cross sections for projectile scattering plus electron emission for electron and positron impact ionization of argon. Solid squares: 100 eV electron impact data from DuBois [30]; open and filled circles: 90 and 120 eV positron impact data of Falke *et al.* [28] placed on an absolute scale as described in the text.

As was seen in the previous figure for elastic scattering, the probabilities for electron and positron impact are very nearly the same in the forward direction. A following figure will show that the contributions in the forward direction are primarily due to the scattered projectile. But, the probabilities differ significantly in the backward direction. Again, for positron impact the cross sections decrease monotonically with angle whereas for electron impact the probabilities reach a minimum value and then increase in the backward direction. Only looking at Fig. (6), the differential data in the forward direction implies an overall higher probability for ionizing interactions for positron impact at 100 eV. However, this is in conflict with the measured values that were shown in Fig. (3).

Fig. (7) clarifies this apparent discrepancy. Here the differential cross sections shown in Fig. (6) have been multiplied by the sine of the scattering (or emission) angle as required for performing an integration over angle. The total area under these curves then corresponds to the total cross sections shown in Fig. (3). As seen, for angles less than 30° the positron and electron impact contributions to the total cross section are nearly the same or perhaps are larger for positron impact. Which is correct depends on cross section information at angles smaller than those measured. But, the major contribution for electron impact is from angles larger than 30°. As a result, the overall electron impact probabilities are significantly larger than those for positron impact. This means that the larger total ionization cross section for electron impact is directly associated with a higher probability that an incoming electron will suffer large angle scattering or that the target electron it ejects will exit the atom at a large angle. Again, referring to Fig. (3), this is consistent with the incoming electron being attracted toward the nuclear charge which enhances the probability that some of these will scatter at larger angles. In addition, the probability that an ejected electron will exit at larger angles for electron impact will be enhanced by the post collision repulsion for electron impact in contrast to the post collision attraction that occurs for positron impact.

For electron impact, it is not possible to examine which of these possibilities is the correct one, or whether both play a role. This is because it is not possible to determine whether the electron observed at a particular angle is the projectile which has been degraded in energy and scattered or whether it is a target electron which has been ejected. However, for positron impact, because of their opposite charges, determining this is possible. A couple decades ago, Falke and coworkers [28] measured how many positrons were scattered at a particular angle. In addition, they measured how many target electrons were ejected at the same angles. The fractions of each are shown in Fig. (8). As seen, only for angles less than approximately 15°, does the probability of scattering dominate; at larger angles, particularly for angles larger than 50°, electron emission is totally dominant. Referring back to Fig. (6), this means that the larger overall ionization cross section for electron impact, as shown in Fig. (3), is due to a higher probability of electron emission in the perpendicular and backward directions by an incoming electron. It should be mentioned that if it were possible to obtain similar information as shown in Fig. (**8**) for electron impact, slight differences in the forward direction will probably be observed. These differences would be due to post collision effects. But, since post collision interactions for electron impact are repulsive, this implies even less contribution in the forward direction by ejected electrons and therefore an even greater contribution at larger angles.

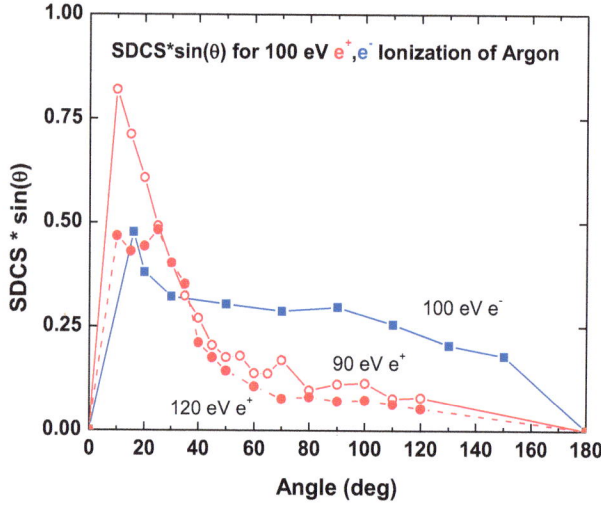

Fig. (7). Singly differential cross section times $\sin(\theta)$ for projectile scattering plus electron emission during ionization of argon by positrons and electrons. Data are same as in Fig. (**5**).

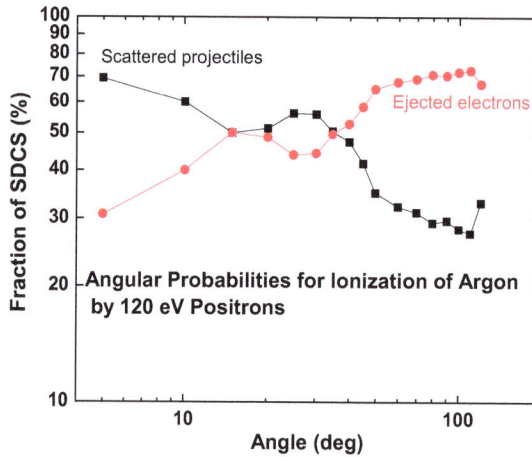

Fig. (8). Fraction of singly differential ionization cross section attributed to scattered projectiles and ejected electrons for 120 eV positron impact on argon. Data of Falke *et al.* [28].

Fig. (**6**) showed that the larger total ionization cross section for 100 eV electron impact is due to larger probabilities of scattering, or emission, into angles larger than approximately 40°. Fig. (**7**) showed that this is most likely due to larger probabilities for electrons being emitted at large angles rather than due to larger scattering probabilities. In the extreme forward direction, Fig. (**8**) showed that scattering, rather than emission, dominates and Fig. (**6**) was inconclusive whether the probabilities were similar for electron and positron or whether they are somewhat larger for positron impact.

These questions are partially answered by looking at available doubly differential cross section data which are shown in Fig. (**9**). Here electron and positron impact data, measured by DuBois and Rudd [31] and Kövér *et al*. [32, 33] are compared.

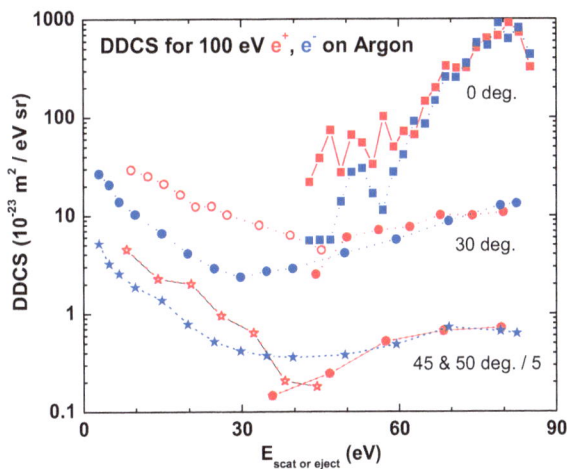

Fig. (9). Doubly differential cross sections for 100 eV electron (blue) and positron (red) impact ionization of argon. Solid squares, 0 degree data of Kövér *et al.* [32]. Open and filled circles, 30 degree positron data of Kövér *et al.* [33] and electron data of DuBois and Rudd [31]. Open and filled stars, 45 degree positron data of Kövér *et al.* [32] and 50 degree electron data of DuBois and Rudd [31], where both sets of data shifted downwards by factor of 5.

The cross sections for projectile scattering plus emission of a target electron for a 100 eV impact energy were measured as functions of scattering/emission angle and energy. First, looking at the positron impact data for observation angles of 30 and 45 degrees where the cross sections for projectile scattering and for target electron emission were independently measured, we see that electron emission dominates for energies less than half the impact energy whereas probabilities for scattering dominate for energies greater than half the impact energy. Second, we note that in both cases, the probabilities for emission by positron impact are larger whereas the probabilities for scattering an incoming positron or electron are nearly identical. Nearly identical scattering probabilities for positrons and electrons were also measured at 0°. These near identical scattering probabilities are in agreement with the singly differential data shown in Fig. (**6**). As for the relative differences for the ejected electron probabilities, at 30° the data in Figs. (**6 and 9**) are in agreement but at 45° they show opposite behaviors. Unfortunately, to obtain an overall picture additional doubly differential data for larger angles is required and such data are currently not available.

V. TRIPLY DIFFERENTIAL CROSS SECTION COMPARISONS

The ultimate comparison that can be made for positron and electron impact involves measuring the complete kinematics for the interaction, *i.e.*, a triply differential measurement where the energy and angle (or the momentum components) are measured for both the scattered positron or electron and the electron that is ejected from the target. Although such measurements were first performed for electron impact many decades ago, the much lower positron beam intensities that can be obtained mean that only a few positron based triply differential studies have been performed. Results of one of these studies performed in our lab are shown in Fig. (**9**). In this study, the overall intensities for binary and recoil interactions were measured for 200 eV positron and electron impact ionization of argon. To remind the reader, binary interactions are two-body interactions involving only the incoming projectile and a single target electron, with all other bound electrons and the target nucleus acting as spectators, whereas in recoil interactions the nuclear charge also plays a role. The reader is also reminded that binary interactions produce a forward electron emission lobe whose direction is primarily determined by momentum transfer while recoil interactions result in large angle, or backward, emission in the same hemisphere that the projectile is scattered in.

Fig. (**10**) shows the ratio of intensities for binary interactions with respect to those for recoil interactions as functions of energy loss and scattering angle for 200 eV positron (left figure) and electron (right figure) impact. As seen, for both positron and electron impact, the binary and recoil interaction probabilities are the same for 0° scattering but binary interactions systematically become more important with respect to recoil interactions as the scattering angle and energy loss increases, *i.e.*, with increasing momentum transfer. More importantly, the relative importance of binary interactions is significantly larger for positron impact. With respect to absolute magnitudes for positron and electron impact, theory predicts that binary interactions are more probable for positron impact compared to electron impact while recoil interactions are less likely [34]. This has been confirmed by experiment [35]. How these triply differential data fit with the singly and doubly differential data shown in Figs. (**7 and 8**) is difficult to say as the singly and doubly differential data involve contributions from a wide range of angles and energies. However, the triply differential data do support the observations made above that electron impact results in a relative enhancement for electron emission in the backward direction.

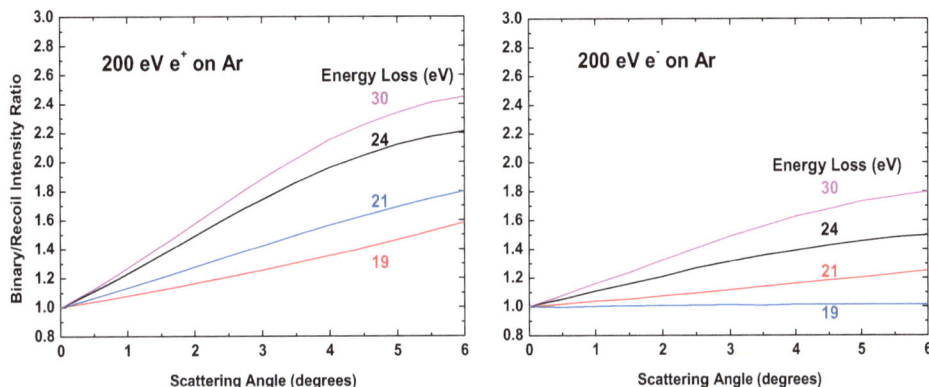

Fig. (10). Ratio of binary to recoil interaction probabilities for 200 eV positron (left) and electron (right) impact single ionization of argon. Ratios are shown for various energy losses. Data are from MST.

CONCLUDING STATEMENTS

We have followed how positron and electron interact with argon atoms from the total (integral) to the highly differential level with the purpose to illustrate similarities and differences and to show how these are related to similar, or different, physical behaviors. In particular, we have discussed how the cross section differences are associated with changes in the incoming particle's trajectory and/or with attraction or repulsion of the target electron cloud, both of which are directly associated with the change of sign of the charge of the incoming particle. We have also tried to show how comparisons of positron and electron impact data assist in our understanding of how matter interacts with matter. Such comparisons also show, to some extent, the similarities and differences of how antimatter interacts with matter. In a general sense, the comparisons provided indicate that in the few hundred eV impact energy range the major difference between incoming positively and negatively charged leptons is an enhanced emission of target electrons for negatively charged projectiles. The major similarity is with respect to projectile scattering which, again in a general sense, is nearly the same for positive and negative charged projectiles.

CONSENT FOR PUBLICATION

Not applicable.

CONFLICT OF INTEREST

The author (editor) declares no conflict of interest, financial or otherwise.

ACKNOWLEDGEMENTS

Declared none.

REFERENCES

[1]   M. Charlton, L. Brun-Nielsen, B.I. Deutch, P. Hvelplund, F.M. Jacobsen, H. Knudsen, G. Laricchiat, and M.R. Poulsen, "Double ionisation of noble gases by positron and electron impact", *J. Phys. At. Mol. Opt. Phys.,* vol. 22, p. 2779, 1989.
[http://dx.doi.org/10.1088/0953-4075/22/17/016]

[2]   L. Chiari, and A. Zecca, "Recent positron-atom cross section measurements and calculations", *Eur. Phys. J. D,* vol. 68, p. 297, 2014.
[http://dx.doi.org/10.1140/epjd/e2014-50436-4]

[3]   M.S. Dababneh, Y.F. Hsieh, W.E. Kauppila, V. Pol, and T.S. Stein, "Total-scattering cross-section measurements for intermediate-energy positrons and electrons colliding with Kr and Xe", *Phys. Rev. A,* vol. 26, p. 1252, 1982.
[http://dx.doi.org/10.1103/PhysRevA.26.1252]

[4]   P.G. Coleman, T.C. Griffith, G.R. Heyland, and T.L. Killeen, Interaction of Low Energy Positrons with Gaseous Atoms and Molecules.*Atomic Physics 4,* G.Z. Putlitz, E.W. Weber, A. Winnacker, Eds., Springer: Boston, MA, 1987.
[http://dx.doi.org/10.1007/978-1-4684-2964-0_19]

[5]   C J Joachain, R Vanderpoortens, and K H Winters, "Optical model theory of elastic electron- and positron-argon scattering at intermediate energies", *J. Phys. B At. Mol. Opt. Phys.,* vol. 10, p. 227, 1977.
[http://dx.doi.org/10.1088/0022-3700/10/2/011]

[6]   P. Van Reeth, M. Szłuińska, and G. Laricchia, "On the normalization of the positron-impact direct ionization cross-section in the noble gases", "Nucl. Instr. and Meth. in Phys", *Res. B,* vol. 192, p. 224, 2002.
[http://dx.doi.org/10.1016/S0168-583X(02)00872-8]

[7]   F.M. Jacobsen, H. Knudsen, U. Mikkelson, and D.M. Schrader, "Single ionization of He, Ne and Ar by positron impact", *J. Phys. B At. Mol. Opt. Phys.,* vol. 28, p. 4691, 1995.
[http://dx.doi.org/10.1088/0953-4075/28/21/016]

[8]   J. Moxom, P. Ashley, and G. Laricchia, "Single ionization by positron impact", *Can. J. Phys.,* vol. 74, p. 367, 1996.
[http://dx.doi.org/10.1139/p96-053]

[9]   S. Mori, and O. Sueoka, "Excitation and ionization cross sections of He, Ne and Ar by positron impact", *J. Phys. B At. Mol. Opt. Phys.,* vol. 27, p. 4349, 1994.
[http://dx.doi.org/10.1088/0953-4075/27/18/028]

[10]   W.E. Kauppila, T.S. Stein, J.H. Smart, M.S. Dababneh, Y.K. Ho, J.P. Downing, and V. Pol, "Measurements of total scattering cross sections for intermediate-energy positrons and electrons colliding with helium, neon, and argon", *Phys. Rev. A,* vol. 24, p. 725, 1981.
[http://dx.doi.org/10.1103/PhysRevA.24.725]

[11]   E. Gargioni, and B. Grosswendt, "Electron scattering from argon: Data evaluation and consistency", *Rev. Mod. Phys.,* vol. 80, p. 451, 2008.
[http://dx.doi.org/10.1103/RevModPhys.80.451]

[12]   H.C. Straub, P. Renault, B.G. Lindsay, K.A. Smith, and R.F. Stebbings, "Absolute partial and total cross sections for electron-impact ionization of argon from threshold to 1000 eV", *Phys. Rev. A,* vol. 52, no. 2, pp. 1115-1124, 1995.
[http://dx.doi.org/10.1103/PhysRevA.52.1115] [PMID: 9912350]

[13]   A.A. Sorokin, L.A. Shmaenok, S.V. Bobashev, B. Möbus, M. Richter, and G. Ulm, "Measurements of electron-impact ionization cross sections of argon, krypton, and xenon by comparison with photoionization", *Phys. Rev. A,* vol. 61, p. 022723, 2000. [http://dx.doi.org/10.1103/PhysRevA.61.022723]

[14]   R.C. Wetzel, F.A. Baiocchi, T.R. Hayes, and R.S. Freund, "Absolute cross sections for electron-impact ionization of the rare-gas atoms by the fast-neutral-beam method", *Phys. Rev. A,* vol. 35, p. 559, 1987. [http://dx.doi.org/10.1103/PhysRevA.35.559]

[15]   D. Rapp, and P. Englander-Golden, "Total Cross Sections for Ionization and Attachment in Gases by Electron Impact. I. Positive Ionization", *J. Chem. Phys.,* vol. 43, p. 1464, 1965. [http://dx.doi.org/10.1063/1.1696957]

[16]   J.C. Gibson, R.J. Gulley, J.P. Sullivan, S.F. Buckman, V. Chan, and P.D. Burrow, "Elastic electron scattering from argon at low incident energies", *J. Phys. B At. Mol. Opt. Phys.,* vol. 29, p. 3177, 1996. [http://dx.doi.org/10.1088/0953-4075/29/14/025]

[17]   I. Iga, L. Mu-Tao, J.C. Nogueira, and R.S. Barbieri, "Elastic differential cross section measurements for electron scattering from Ar and O2 in the intermediate-energy range", *J. Phys. B At. Mol. Opt. Phys.,* vol. 20, p. 1095, 1987. [http://dx.doi.org/10.1088/0022-3700/20/5/025]

[18]   R. Panajotović, D. Filipović, B. Marinković, V. Pejčev, M. Kurepa, J. Vušković, and B. Phys, "Critical minima in elastic electron scattering by argon", *J. Phys. B At. Mol. Opt. Phys.,* vol. 30, p. 5877, 1997. [http://dx.doi.org/10.1088/0953-4075/30/24/022]

[19]   S K Srivastava, H Tanaka, A Chutjian, and S Trajmar, "Elastic scattering of intermediate-energy electrons by Ar and Kr", *Phys. Rev. A.,* vol. 23, p. 2156, 1981. https://journals-ap--org.libproxy.mst.edu/pra/abstract/10.1103/PhysRevA.23.2156

[20]   J.E. Furst, D.E. Golden, M. Mahgerefteh, J. Zhou, and D. Mueller, "Absolute low-energy e− -Ar scattering cross sections", *Phys. Rev. A,* vol. 40, p. 5592, 1989. [http://dx.doi.org/10.1103/PhysRevA.40.5592]

[21]   R.D. DuBois, and M.E. Rudd, "Differential cross sections for elastic scattering of electrons from argon, neon, nitrogen and carbon monoxide", *J. Phys. B At. Mol. Opt. Phys.,* vol. 9, p. 2657, 1976. [http://dx.doi.org/10.1088/0022-3700/9/15/016]

[22]   H. Bluhme, H. Knudsen, J.P. Merrison, and K.A. Nielsen, "Ionization of argon and krypton by positron impact", *J. Phys. B At. Mol. Opt. Phys.,* vol. 32, p. 5835, 1999. [http://dx.doi.org/10.1088/0953-4075/32/24/317]

[23]   P. McCallion, M.B. Shah, and H.B. Gilbody, "A crossed beam study of the multiple ionization of argon by electron impact", *J. Phys. B At. Mol. Opt. Phys.,* vol. 25, p. 1061, 1992. [http://dx.doi.org/10.1088/0953-4075/25/5/017]

[24]   R. Rejoub, B.G. Lindsay, and R.F. Stebbings, "Determination of the absolute partial and total cross sections for electron-impact ionization of the rare gases", *Phys. Rev. A,* vol. 65, p. 042713, 2002. [http://dx.doi.org/10.1103/PhysRevA.65.042713]

[25]   J.F. Williams, and B.A. Willis, "The scattering of electrons from inert gases. I. Absolute differential elastic cross sections for argon atoms", *J. Phys. B At. Mol. Opt. Phys.,* vol. 8, p. 1670, 1975. [http://dx.doi.org/10.1088/0022-3700/8/10/017]

[26]   R.H.J. Jansen, F.J. de Heer, H.J. Luyken, B. van Wingerden, and H.J. Blaauw, "Absolute differential cross sections for elastic scattering of electrons by helium, neon, argon and molecular nitrogen", *J. Phys. B At. Mol. Opt. Phys.,* vol. 9, p. 185, 1976. [http://dx.doi.org/10.1088/0022-3700/9/2/009]

[27]   L. Dou, W.E. Kauppila, C.K. Kwan, and T.S. Stein, "Observation of structure in intermediate-energy positron-argon differential elastic scattering", *Phys. Rev. Lett.,* vol. 68, no. 19, pp. 2913-2916, 1992. [http://dx.doi.org/10.1103/PhysRevLett.68.2913] [PMID: 10045527]

[28] T. Falke, T. Brandt, O. Kuhl, W. Raith, and M. Weber, "Differential Ps-formation and impact-ionization cross sections for positron scattering on Ar and Kr atoms", *J. Phys. B At. Mol. Opt. Phys.,* vol. 30, p. 3247, 1997.
[http://dx.doi.org/10.1088/0953-4075/30/14/015]

[29] Á. Kövér, G. Laricchia, and M. Charlton, "Doubly differential cross sections for collisions of 100 eV positrons and electrons with argon atoms", *J. Phys. B At. Mol. Opt. Phys.,* vol. 27, p. 2409, 1994.
[http://dx.doi.org/10.1088/0953-4075/27/11/031]

[30] R.D. DuBois, *"Absolute Differential Cross Sections for 20-800 eV e- -Ar and e- -N2 Collisions for Elastic Scattering and Secondary Electron Production"*, Ph. D. Thesis, University of Nebraska-Lincoln, 1975.

[31] R.D. DuBois, and M.E. Rudd, "Absolute doubly differential cross sections for ejection of secondary electrons from gases by electron impact. II. 100-500-eV electrons on neon, argon, molecular hydrogen, and molecular nitrogen", *Phys. Rev. A,* vol. 17, p. 843, 1978.
[http://dx.doi.org/10.1103/PhysRevA.17.843]

[32] Á. Kövér, G. Laricchia, and M. Charlton, "Ionization by positrons and electrons of Ar at zero degrees", *J. Phys. B At. Mol. Opt. Phys.,* vol. 26, p. L575, 1993.
[http://dx.doi.org/10.1088/0953-4075/26/17/008]

[33] Á. Kövér, G. Laricchia, and M. Charlton, "Doubly differential cross sections for collisions of 100 eV positrons and electrons with argon atoms", *J. Phys. B At. Mol. Opt. Phys.,* vol. 27, p. 2409, 1994.
[http://dx.doi.org/10.1088/0953-4075/27/11/031]

[34] S. Sharma, and M.K. Srivastava, "Triple-differential cross sections for the electron- and positron-impact ionization of helium in an improved second Born approximation", *Phys. Rev. A,* vol. 38, p. 1083, 1988.
[http://dx.doi.org/10.1103/PhysRevA.38.1083]

[35] O.G. de Lucio, S. Otranto, R.E. Olson, and R.D. DuBois, "Triply Differential Single Ionization of Argon: Charge Effects for Positron and Electron Impact", *Phys. Rev. Lett.,* vol. 104, p. 163201, 2010.
[http://dx.doi.org/10.1103/PhysRevLett.104.163201] [PMID: 20482046]

CHAPTER 2

Translational Nascent Kinetic Energy of the CH_2Cl_2 Molecule After Photoexcitation Around Cl 2p Edge and its Implications to the Physics of Atmosphere

K.F. Alcantara[1], A.B. Rocha[3], A.H.A. Gomes[4], W. Wolff[2], L. Sigaud[5] and A.C.F. Santos[2,*]

[1] *Instituto Nacional de Tecnologia, 20081-312, Rio de Janeiro, RJ, Brazil*

[2] *Instituto de Física, Universidade Federal do Rio de Janeiro-21941-972. Rio de Janeiro, RJ, Brazil*

[3] *Instituto de Química, Universidade Federal do Rio de Janeiro-21941-909. Rio de Janeiro, RJ, Brazil*

[4] *Instituto de Física Gleb Wataghin, Universidade Estadual de Campinas-13083-859. Campinas, SP, Brazil*

[5] *Instituto de Física, Universidade Federal Fluminense-24210-346. Niterói, RJ, Brazil*

Abstract: The kinetic energy release (KER) of molecular fragments is of major interest in molecular reaction dynamics. When dissociation reactions of polyatomic ions occur, some of the excess internal energy of the ion is released as kinetic energy of the two fragments. The KER provides important information about the structures of the molecular species involved and on the energetics and dynamics of the reaction. To meet this end, the translational kinetic energy release distribution spectra of selected ionic fragments, produced through dissociative single and double photoionization of dichloromethane (DCM) molecule at photon energies around the chlorine L edge, were measured from the shape and width of the experimentally obtained time-of-flight (TOF) distributions. The kinetic energy release distributions (KERD) spectra exhibit either smooth profiles or structures, depending on the ionic fragment and photon wavelength. In general, the heavier the ionic fragments, the lower are their average KERDs. In contrast, the light H^+ fragments are observed with kinetic energies centered around 4.5-5.5 eV, depending on the photon wavelength. It was noticed that the change in the photon wavelength involves a change in the KERDs, pointing out different processes or transitions taking place in the break-up process. In the particular case of double ionization with the ejection of two charged fragments, the kinetic energy distri-

---

* **Corresponding author A. C. F. Santos:** Instituto de Física, Universidade Federal do Rio de Janeiro - 21941-972. Rio de Janeiro, RJ, Brazil; Tel: +55 21 3938-7947; Fax: +55 21 3938-7368; E-mail: toni@if.ufrj.br

butions present own characteristics compatible with the Coulombic fragmentation model. Intending to interpret the experimental data singlet and triplet states at the chlorine L edge of the dichloromethane molecule, associated to the Cl ($2p \rightarrow 10a_1^*$) and Cl ($2p \rightarrow 4b_1^*$) transitions, were determined at multiconfigurational self-consistent field (MCSCF) level and multi reference configuration interaction (MRCI). These states were selected to form the spin-orbit coupling matrix elements, which after diagonalization results in a spin-orbit manifold. Minimum energy pathways for dissociation of the molecule were additionally calculated aiming to give support to the presence of the ultra-fast dissociation mechanism in the molecular break-up.

Keywords: Fragmentation, PEPICO, PEPIPICO, Kinetic energy release, Time-of-flight spectrometer, MCSCF, Spin-orbit, Cl 2p, Translational energy, Photoions, Shallow core, Photoexcitation, Photoionization, Dichloromethane, CH_2Cl_2.

I. INTRODUCTION

Chemical processes involving single and double ionized ions are of main importance in shaping the equilibrium distribution of cations in planetary ionospheres. Photoionization and charged particle impact ionization of the atmospheric components are responsible for the initial generation of the electron-ion pairs. Consequently, a systematic knowledge of the guiding chemical processes is required for an accurate understanding of ionospheric composition and behavior. The ionization threshold for the principal atmospheric components, in atomic and molecular form, is in the range 8-12 eV.

Solar radiation in the solar extreme ultraviolet (EUV) and x-ray range of energy excites, dissociates, and ionizes the neutral components in the upper part of the atmosphere. The precipitations of solar EUV radiation and charged particles, mainly electrons, are the two main causes of energy transport into the Earth's ionosphere. Relatively long wavelength photons ($\lambda < 91$ nm) may possibly cause ionization. The exact distribution of the ionization products depends on the energy dependent cross sections as well as the atmospheric gases. The unambiguos distribution of the manner in which the energy flows through the system is very important in understanding the composition and thermal structure of the ionospheric plasmas.

Photoionization of the neutral molecular constituents in Earth's atmosphere creates free electron-ion pairs, being the major cause of ionization in the ionosphere. The energy of the ionizing photons is beyond, in some cases, the threshold ionization energy. The excess of energy goes either into the electron kinetic energy, or excitation of the produced ion. The reason why the electrons pick up most of the available kinetic energy is that the ions are much heavier than the electrons, and consequently, the ions acquire very little, but significant, recoil energy during the photoionization process due to the role of internal forces.

The portrayal of the chemistry of atmospheric media strongly depends on ion–molecule reaction rate data. An important parameter that controls the ion-molecule reactions rates for Earth`s ionospheric chemistry modeling, is the temperature/collision energy.

The ozone layer, the main absorber of dangerous solar UV radiation in the Earth`s atmosphere, is a vital shield for the mankind. Thus, the large rate of dissipation of chlorine containing molecules released by human action into the atmosphere presents a major threat to the ozone layer. The EUV and x-ray sunlight radiation induced molecular ionization and fragmentation produces chlorine radicals, which are understood to be the main cause for damaging the atmospheric ozone.

Studies involving stratospheric chlorine have been performed since the initial work of M. J. Molina and F. S. Rowland. They proposed that atomic chlorine released from chlorofluorocarbons in the stratosphere would favor the catalytic destruction of the ozone layer. Some stratospheric ozone models have been considerably expanded since the 1970s. In addition, the role played by the various chlorine compounds engaged in the process has been evaluated by a large number of studies. Nevertheless, the initial work by Molina, Rowland, and Crutzen on the role of anthropogenic chlorine in stratospheric ozone was rewarded a Nobel Prize in Chemistry in 1995 presented to for their contribution to the knowledge of the role of nitrogen species in the destruction of ozone.

Short and long-lived chlorine-containing organic molecules are freed on the Earth's surface. The chlorofluorocarbons begun to be manufactured in the 1930s. They are manmade, being used mostly as refrigerants and propellants. These chlorine-containing gases are released into the stratosphere, and, since they are virtually inert in the troposphere and lower stratosphere, they endure the transport into the middle stratosphere. Because they are the sources of stratospheric chlorine, these gases are known as 'chlorine source gases'.

Catalytic cycles concerning chlorine compounds have been suggested in the literature. Molina and Rowland (1974) identified halogenated hydrocarbons as an source of stratospheric chlorine and postulated catalytic ozone destruction. The catalytic cycle

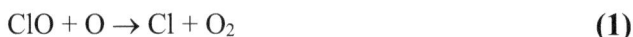

$$ClO + O \rightarrow Cl + O_2 \qquad\qquad \textbf{(1)}$$

$$Cl + O \rightarrow ClO + O_2 \tag{2}$$

Which the net result is $\quad O + O_3 \rightarrow 2O_2$ (3)

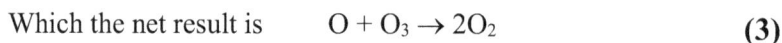

The chance of the ozone layer which protects life on Earth from dangerous solar ultraviolet radiation being destroyed, called the attention of scientists in stratospheric chemistry, as well as the willingness of the representatives to invest money on these studies.

After the absorption of a sufficiently high-energy photon by the CH_2Cl_2 the excess of energy of the dissociating molecule is spread mostly among the fragments rather than inducing the fluorescence emission because the molecule is made by light atoms (H, C and Cl). The excited molecular states prompted by the photo-absorption can follow several fragmentation pathways producing fragments with distinct translational energies following the rules of the fragmentation dynamics, obeying momentum and energy conservation laws. In this perspective, KER analysis, constituting a probe of the molecular fragmentation of the charged system, covers this lack to a large extent, and it is of basic interest for understanding the structures of the moieties concerning the fragmentation dynamics, provides information on the excited states, on the character of repulsive states in the fragmentation process and insights on the electronic states and potential energy functions [1 - 3].

The scope of this work is to obtain the average kinetic energies and KERD of several ionic fragments resulting from the dissociative photoionization of the singly and doubly charged CH_2Cl_2 molecule by time-of-flight spectra. It is well known that in time-of-flight (TOF) mass spectra the peak width of the parent molecular ion is composed by the thermal velocities and spatial distribution of the target species, convoluted with an instrumental broadening due to electronics [4]. On the other hand, for the break up channels, ions resulting from molecular fragmentation depict mostly TOF peak profiles broadened by the changeover of the dissociation energy into kinetic energy release (KER) and the momentum distribution between a two and a three body break up. It should be pointed out that although the energy distributions of the charged fragments can be determined, the total energy distribution cannot to be ascertained when neutral fragments are involved in the break-up channels carrying part of it. Notwithstanding, KERDs of molecular fragments have to be interpreted with caution due to the internal energy of the fragments.

The selected molecule in this study, Dichloromethane (DCM), is amply employed as a paint stripper, degreaser and to decaffeinate coffee and tea. Its high volatility is useful as an aerosol spray propellant as well as a blowing agent for polyurethane foams [5 - 7]. The fragmentation of DCM induced by solar wind and cosmic rays (photons and charged particles) has been related to the depletion of the ozone layer [5 - 7]. The main mechanism related to the depletion of ozone is *via* reactions with chlorine atoms. Chlorine compounds can undergo dissociation, releasing a chlorine atom as the result of a collision process [8, 9]. Atomic chlorine is highly reactive and breaks ozone molecules, forming a chlorine monoxide and an oxygen molecule. Chlorine monoxide, in turn, can react with either an oxygen atom or another ozone molecule, resulting in a chlorine atom and one or two oxygen molecules, respectively, leading to a feedback reaction mechanism where a single Chlorine atom can break down a large number of ozone molecules.

Measurements of DCM in the North Atlantic and Labrador Sea show that its distribution closely matches that of CFC 11, which is acknowledged to have an entirely atmospheric source. Although DCM has a short lifetime in the atmosphere, it endures from years to decades in the intermediate and deep ocean. Reports of an ocean source of DCM, inferred from measured super-saturation, may reflect material that had an atmospheric source with net uptake during winter and release in summer.

II. EXPERIMENT

The experimental apparatus has been reported previously [8 - 10], and only a brief description is included here. The CH_2Cl_2 molecule was photoionized by soft X-rays from the LNLS synchrotron radiation facility monochromatized in the TGM beam line. The photoionization products of CH_2Cl_2 were separated according to their mass-to-charge ratio by a time-of-flight (TOF) mass spectrometer, using the ejected electrons and the cations as start and stop signals, respectively, for a time to digital converter (TDC).

An important concept in time-of-flight mass spectrometry that should be comprehended is its mass resolution. Mass resolution is the experimental observable of the resolving power of the spectrometer. The resolving power of a time-of-flight spectrometer is its ability to separate fragments according to their mass. The time-of-flight, $t(m/q)$, of a specific fragment depends on its mass-to-charge ratio (m/q), *i.e.*

$$t(m/q) \propto \sqrt{\frac{m}{q}} \qquad \qquad \textbf{(4)}$$

From equation (1), one obtains:

$$dt \propto \frac{1}{2}\sqrt{\frac{m}{q}}\frac{dm}{m}$$

and

$$\frac{dt}{t} = \frac{dm}{2m} \tag{5}$$

or

$$\frac{\Delta m}{m} = \frac{2\Delta t}{t}$$

The relative resolution, $(\Delta m / m)$, observed in time-of-flight mass spectrometry, can be calculated from the peaks observed and is derived from Eq. (4).

From these TOF profiles, the KERD of several cations were deduced for several photon energies. The extraction field chosen for the spectrometer in the experiment is high enough (70000 V/m) to guarantee 100% collection efficiency for ions up to 30 eV [11]. The experiment was performed using an effusive gas jet at room temperature. Although the arising thermal energies might influence the TOF profiles, they are a minor factor compared to the released energies. The TOF distribution of the singly ionized parent molecules is described by a Maxwell-Boltzmann energy profile, which was also checked measuring argon gas spectra at room temperature.

Consider the dissociation of a molecule M into a charged fragment of mass m_1 and a neutral fragment of mass m_2 after the photoionization of the parent molecule M, which can be represented by the equation:

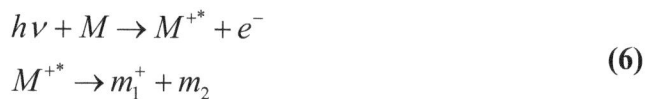

$$h\nu + M \rightarrow M^{+*} + e^-$$
$$M^{+*} \rightarrow m_1^+ + m_2 \tag{6}$$

The fragment ion m_1^+ is formed with kinetic energy U_1 and velocity $v_1 = (2U_1/m_1)^{1/2}$. The excess internal energy of the parent ion, M^{+*}, is shared among all the internal degrees of freedom. Part of this internal energy is released in the relative translation of departing fragments. This energy is then called the kinetic energy release (KER).

The KERD of the CH_2Cl_2 fragments were investigated in the energy range around the chlorine L edge (195 eV- 215 eV) by means of the analysis of the TOF profiles through the PEnPICO technique. The mean kinetic energy release (KER) in the fragmentation process gives rise to a velocity spread of the fragment ions. This spread can be tracked by measuring the TOF distributions of the fragments. The time-of-flight peak profiles allow us to compute the average kinetic energy release measurements [4, 11] of the coincidence products of the CH_2Cl_2 molecule. In other words, from the time-of-flight mass spectroscopy one can easily obtain information about the initial energy of the ions that is carried by the width and shape of mass peak profiles. If a fragment of mass m and charge q has been generated in a constant homogeneous electric field with intensity E with kinetic energy U, its time of flight is given by

$$T(^m/_q, U) = k\sqrt{\frac{m}{q}} \pm \Delta T(U, \theta) \tag{7}$$

$$\Delta T(U, \theta) = \sqrt{2mU}\,\frac{\cos\theta}{qE} \tag{8}$$

Where k is a constant depending on the spectrometer dimensions and the applied voltages, which can be determined by a linear regression to the peak maxima of known ions, and θ represents the angle between the initial momentum $p = \sqrt{2mU}$ and the direction of the electric field. In the case of isotropic distribution and single-valued kinetic energy U, that the mass line is a rectangle of full width W given by

$$W = 2\Delta T_{av} = 2\frac{\sqrt{2mU}}{qE} \tag{9}$$

Since the fragments are utterly space focused and the electric field in the interaction region is uniform, the mean KER in the fragmentation process, $\langle U \rangle$, can be calculated from the peak FWHM through the formula [4, 11]

$$\langle U \rangle = \left(\frac{qE\Delta t_{FWHM}}{2}\right)^2 \frac{1}{2m} \tag{10}$$

where q is the ion charge state, E is the electric field in the extraction region, Δt_{FWHM} is the peak width (FWHM), and m is the ion mass. The FWHM provides a very good representation of the *mean or average* kinetic energy. This is due since the FWHM embraces the advantage of being quite indifferent to statistical fluctuations. Another advantage resides in the fact that the FWHM can precise and straightforwardly revised for broadening due to instrumental and thermal causes. Thus, the method of obtaining kinetic energies from peak profiles of conventional time-of-flight spectrum constitutes a very simple procedure, despite of its restrictions.

The zero of the KERD scale is obtained from the position (line centroids) of the fragment TOF profiles, assuming a symmetrical profile. Because the KERD is proportional to the square of Δt, the absolute precision of the energies, δU, is better for low KER ($\delta U/U = 2\delta\Delta t/\Delta t$), where $\delta\Delta t$ is the time resolution.

Let us assume that the initial momenta of the fragments are isotropically distributed and that the detection of ions is free from dead-time effects and from discrimination with respect to the magnitude or direction of the momenta. Thus, the peak profiles are symmetric. Then, one reflects the left half of the observed peak profile at the centre and sum it to the right one, giving rise to a half-peak shape function $I(t)$, which is normalized to unity, dividing it by the total number N of counts in the peak. If the energy is single valued, then

$$I(t) = \left(\frac{W}{2}\right)^{-1} = \left(\frac{\sqrt{2mU}}{qE}\right)^{-1}$$

. In reality the KER underlies a distribution $N(U)$, whence $I(t)$ becomes a superposition of such rectangles, as shown in Fig. (**1**).

The kinetic energy distributions $N(U)$ can be worked out from the shape of the lines $I(t)$ observed in the TOFMS, through the relation

$$N(U) = \frac{2m}{(qE)^2}\frac{dI(t)}{dt} \qquad (11)$$

In the case of the PE2PICO coincidences, the corresponding average KER and KERDs were extracted after projecting the coincidences islands onto the T_1 and T_2 axi [10], which correspond to the time-of-flight of the first and second fragments to hit the detector, respectively. KER and KERD in fragmentations of singly and multiply charged ions provide information concerning ion reaction energetics and dynamics.

III. THEORETICAL APPROACH

For atomic systems, core level spectroscopy uses the terms nl_j, being n is the principal quantum number, l the angular momentum quantum number and $j = l + s$, where s is the spin angular momentum number. Each orbital level excluding the s levels ($l = 0$) exhibits a doublet with the two possible states with distinct binding energies. This is the well known spin-orbit splitting, also known as j-j coupling. The peaks exhibit specific area ratios depending on the degeneracy of the spin state, namely, the number of distinct spin combinations that can give rise to the total j ($|l + s|$, $l + s-1|$, ..., $|l-s|$). For the $2p$ spectra, where n is 2 and l is 1, j will be 1/2 and 3/2. The area ratio for the two spin orbit peaks ($2p_{1/2}$:$2p_{3/2}$) will be 1:2 (2 electrons in the $2p_{1/2}$ level and 4 electrons in the $2p_{3/2}$ level).

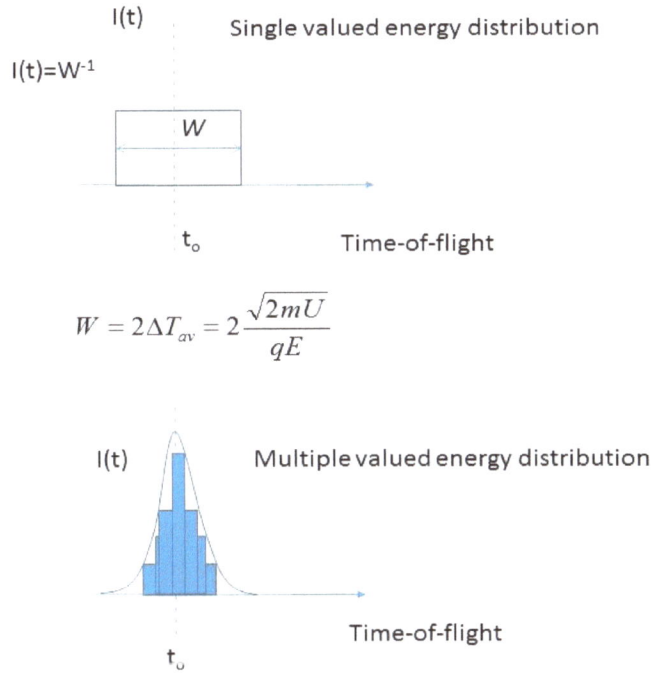

$$W = 2\Delta T_{av} = 2\frac{\sqrt{2mU}}{qE}$$

Fig. (1). Single (top) and multiple valued energy distributions.

In order to designate the transitions which are responsible to the structures observed in the experimental results (the combined experimental and theoretical investigation is presented in detail in the results discussion section), high-level ab initio quantum mechanical calculations were performed taking into account the spin-orbit coupling. Singlet and triplet states at Cl 2p edge of the CH_2Cl_2 molecule, corresponding to the Cl ($2p \rightarrow 10a_1^*$) and Cl ($2p \rightarrow 4b_1^*$) transitions, were calculated in order to form a basis set of molecular states from which spin-orbit splitting can be inferred. The method used was described previously [12]. Initially, a multiconfigurational self-consistent field (MCSCF) calculation is performed in order to obtain orbitals averaged among ground and L-core states, followed by multireference configuration interaction (MRCI) in order to establish a set of singlet and triplet states at the L excitation edge. Then, the full Breit-Pauli Hamiltonian is formed and diagonalized on this state basis. The active space in the MCSCF step is composed of inner-shell L orbitals of chlorine atoms and the corresponding unoccupied orbital in each case ($10a_1$* and $4b_1$*) in a state-averaged MCSCF, by considering only singly-excited configurations for each state. This is a very simple wave function but it is able to take into account the most important effect of an inner-shell state, that is, the relaxation of the orbitals due to the formation of a core hole. The correlation effects are recovered by MRCI. With these eigenfunctions of the electronic Hamiltonian, H_{el}, the total Hamiltonian, $H_{el} + H_{SO}$, is diagonalized, where H_{SO} is the spin-orbit Hamiltonian in the full Breit-Pauli form [13 - 16]. Finally, the transition moment involving the ground state and the states of the spin-orbit manifold are calculated in order to determine the relative transition intensities. All calculations were done with the Molpro package [17] with Dunning's aug-cc-pVTZ-DK basis set. Scalar relativistic effects are taken into account by the Douglas-Kroll-Hess Hamiltonian up to third order in all steps of the calculation.

Minimum energy pathways for dissociation of the molecule were additionally calculated aiming to give support to the presence of the ultra-fast dissociation mechanism in the molecular break-up. Potential energy curves were calculated, for some representative states, as a function of the C-Cl distance. For a fixed C-Cl distance all other coordinates are optimized, meaning that this is a minimum energy path. For inner-shell states, such as Cl ($2p \rightarrow 10a_1$*), the inner-shell orbital is relaxed in the first step and then kept fixed for all points of the potential curve. This is done to avoid the variational collapse to a low-lying electronic state. All other orbitals are relaxed for each geometry optimization. For the ground state calculation, all orbitals are relaxed without restriction, since the problem of variational collapse does not apply. For these potential curve calculations, a smaller basis set was used, *i.e.*, cc-pVDZ and spin-orbit effects were not taken into account.

IV. RESULTS AND DISCUSSION

Fig. (**2**) shows the total ion yield spectrum of the CH_2Cl_2 molecule around the Cl L edge. Included in this figure are TOF profiles of the ionic isotopic chlorine atoms, $^{35}Cl^+$ and $^{37}Cl^+$, at selected photon energies. These profiles were chosen to point out that changing the photon wavelength implies in a concomitant change in their shape and their width and, consequently, a change in the KERDs. The doublet structures in the Total Ion Yield (TIY) spectrum, at 200.7 eV (band A) and 202.3 (band B), are due to the spin-orbital (SO) splitting of the $2p_{3/2}$ and $2p_{1/2}$ levels of chlorine. Spin-orbit effects generally become apparent as energy splitting among states that exhibit similar electronic configuration. Even though molecular-field effects are characteristic of the chemical environment of the observed electron, molecular core levels are usually treated as atomic-like. These molecular-field effectiveness produce, among other outcomes, energy shifting in the core-level and valence orbitals. Thus, the trend to treat molecular inner shells as essentially atomic-like is justified by the fact that molecular-field properties induce negligible perturbations on core-level spectra.

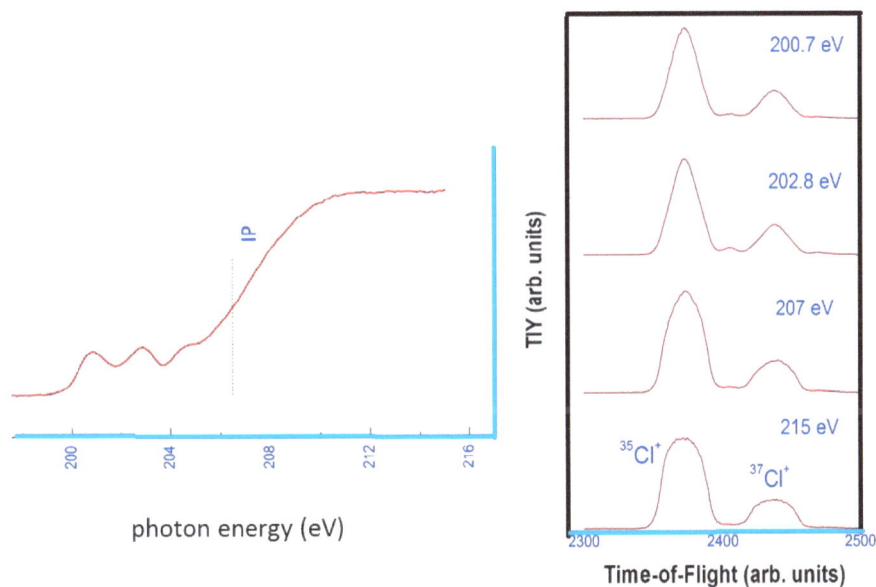

Fig. (2). Left side: Total ion yield (TIY) of the CH_2Cl_2 molecule as a function of the photon wavelength. Right side: Time-of-flight spectra of ^{35}Cl e ^{37}Cl at selected photon energies. See text for the peak designations (A-B) and dot lines explanations.

In Fig. (**3**), band A is here tentatively assigned to the transition from the Cl ($2p_{3/2}$ → $10a_1^*$ - 200.7 eV) [18, 19], while band B, to the Cl ($2p_{1/2}$ → $10a_1^*$ - 202.0 eV) and ($2p_{3/2}$ → $4b_1^*$ - 202.8 eV) transitions. The two ionization limits due to the spin-orbit splitting of the L hole, Cl $2p_{3/2,1/2}$ → ∞ (IP) are designated by the dot lines at 206.4 eV and 208.0 eV [18, 19], respectively.

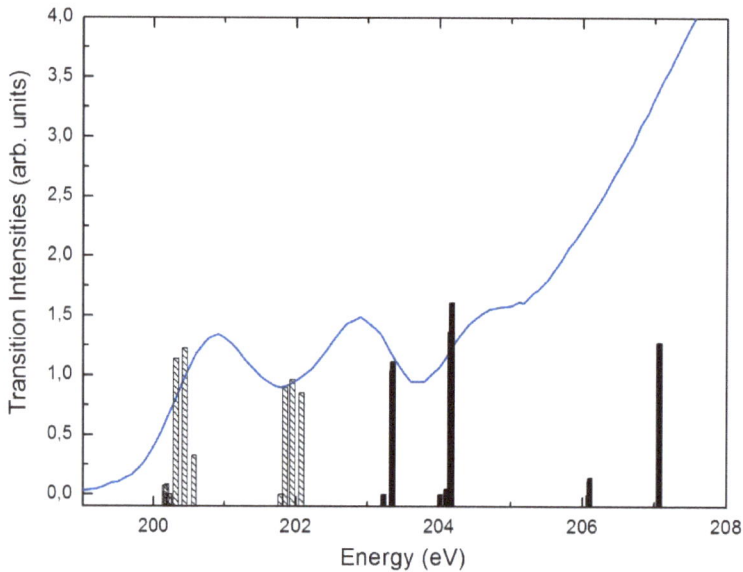

Fig. (3). Transition intensities as a function of the photon wavelength. Full line: experimental data; dashed bars: theoretical calculation for the 2p →10a$_1$* transition; full bars: theoretical calculation for the L → 4b$_1$* transition.

In order to verify whether the above assignments are consistent and to shed some light on the splitting of the spin-orbit manifold we have performed theoretical calculations. In Fig. (**2**) the theoretical calculations and experimental results of the CH$_2$Cl$_2$ molecule around the Cl L edge are compared. The total ion yield, after the subtraction of an offset due to contributions from valence shell ionization, has been normalized in order to compare with the theoretical transition intensities corresponding to the Cl (L → 10a$_1$*) and Cl (L → 4b$_1$*) transitions. Dashed bars stand for theoretical results for the L →10a$_1$* transition, where spin-orbit effects were taken into account as explained above, whilst full bars stand for theoretical calculations for the L → 4b$_1$* transition. The ground state plus 24 L-Shell states were used to form the SO matrix. Therefore, there are 24 transitions from the ground to the excited states of the spin-orbit manifold. In some cases, transition intensity vanishes. The main contributions are shown in Fig. (**3**). The first transition lies at 200.2 eV and the last lies slightly above 207 eV, *i.e.*, inside the continuum of ionization of the L edge. If more states were used to form the SO matrix, the resulting states would lie still higher inside the continuum.

As the theoretical quantities are calculated as expectation values (single values), theoretical line widths cannot be indicated in the theoretical spectrum. Notwithstanding, calculations can highlight new information. For instance, for the

L → 10a$_1$* transition, L$_{3/2}$ components lie around 200.5 eV and L$_{1/2}$ components lie around 202 eV. There are components, instead of a single absorption, due to molecular splitting. If we take the center of the components as reference, the spin-orbit splitting is calculated as 1.51 eV, which is consistent, for instance, with the value calculated and measured for HCl[12]. It is worth mentioning that the experimental spectrum cannot assign the spin-orbit splitting since there is not enough resolution. All states resulting from the L → 10a$_1$* manifold lie inside the unresolved band A. So, when the natural line width and the limitation of resolution are taken into account, not only the Cl (L$_{3/2}$ → 10a$_1$*) but also the Cl (L$_{1/2}$ → 10a$_1$*) transitions contribute to band A, which was not considered in the initial experimental assignment. Additionally, it can be seen that Cl (L$_{1/2}$ → 10a$_1$*) components should also contribute to the band B, as well as Cl (L$_{3/2}$ → 4b$_1$*) confirming the initial experimental assignment.

IV.1. Pepico Spectra

Consider that a singly charged parent molecular ion m$^+$ of mass m, initially at rest dissociates into one singly charged fragment and one or more neutral fragments:

$$m^+ \rightarrow m_1^+ + m_2 + m_3 +. \tag{12}$$

The excess internal energy ε of the parent ion is distributed among the vibrational, rotational, (internal) and translational (external) degrees of freedom of the departing fragments. Applying the laws of energy and momentum conservation:

$$\varepsilon = \sum_i \varepsilon_i$$
$$0 = \sum_i \vec{p}_i \tag{13}$$

Where $\varepsilon_i = \dfrac{\vec{p}_i^{\,2}}{2m_i}$ are the kinetic energy release of the fragments and $\mathbf{p}_i$ are their momenta. In the present set-up, only the KER of the charged fragment can be measured.

In the range between 195 and 215 eV, the most abundant fragments in the mass spectra are Cl$^+$ (31 – 44%), H$^+$ (13-14%) and CH$_2^+$ (9- 16%), as listed in a previous work [19] and shown in Fig. (**4**), as an example. Other break-ups have been detected, but with relative intensities of less than 4%.

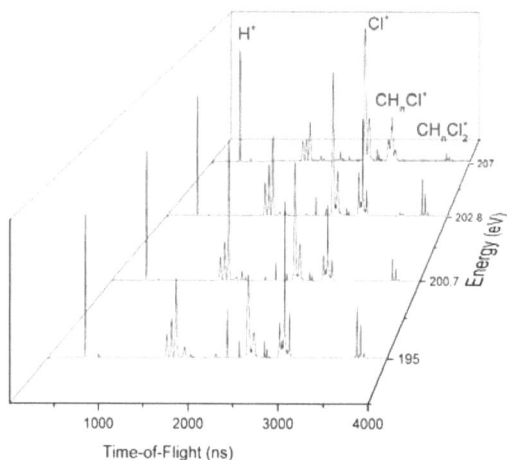

Fig. (4). PEPICO spectra measured around the Cl L edge.

The KERD of these fragments and of the C^+ fragment, corresponding to the full atomization of the molecule, were derived from the peak profiles of the TOF spectra. Figs. (**5-10**) show the average KER (FWHM) of selected photoions of the valence and core-excited CH_2Cl_2 molecule as a function of the incident photon wavelength.

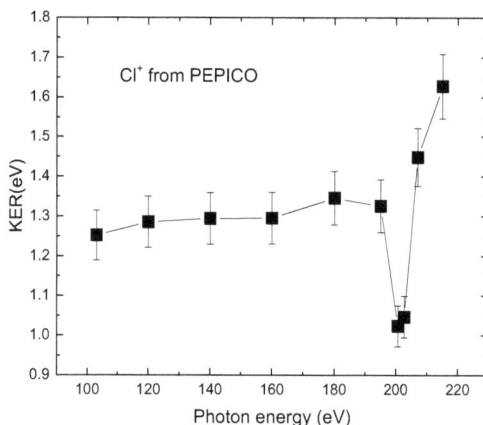

Fig. (5). Average center-of-mass KER (FWHM) of Cl^+ fragment from the CH_2Cl_2 molecule as a function of the incident photon wavelength. The lines at 200.7 eV ($L_{3/2} \rightarrow 10a_1^*$) and 202.8 eV ($L_{3/2} \rightarrow 4b_1^*$) indicate the energy transitions. It was observed that the FWHM TOF profiles are narrower at the Cl ($L_{3/2} \rightarrow 10a_1^*$) resonance. Because the mass of the photoelectron is very small in comparison to the mass of the ion, the whole energy available in the process is essentially converted to electron kinetic energy. The fragments exhibit a considerable increase in the KER for excitation to molecular orbitals near the Cl $L_{3/2,1/2} \rightarrow \infty$ (IP), which have a strong Rydberg character.

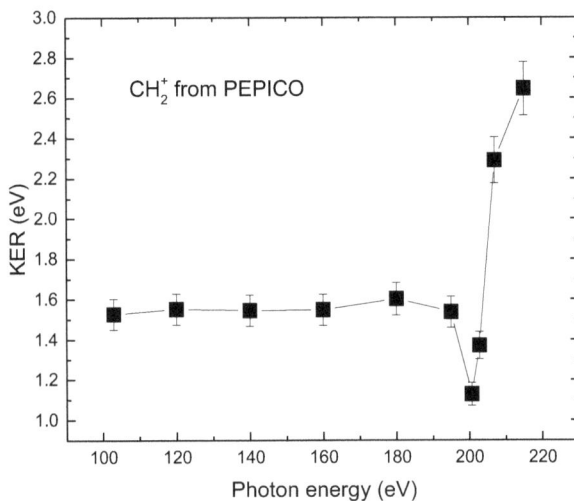

Fig. (6). Average center-of-mass KER (FWHM) of CH_2^+ fragment from the CH_2Cl_2 molecule as a function of the incident photon wavelength.

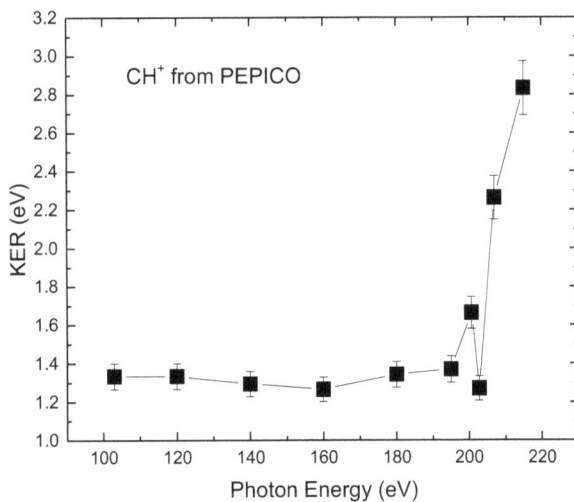

Fig. (7). Average center-of-mass KER (FWHM) of CH^+ fragment from the CH_2Cl_2 molecule as a function of the incident photon wavelength.

The CH_2^+ yield presents a maximum at the Cl ($2p_{3/2} \rightarrow 10a_1^*$) resonance, which was interpreted by Lu *et al.*[18] as due to a fast dissociation through a highly repulsive potential curve. The amount of energy released to translation of products depends on the details of the potential energy surface. Here we present calculations of the potential curves in order to give support to the assertion of a fast dissociation. Fig. (**11**) shows the profile of energy *versus* C-Cl distance for the inner-shell singlet Cl ($2p \rightarrow 10a_1{}^*$) neutral state. It is quite evident the curve is repulsive and can promote fast dissociation. Such process was suggested for methyl chloride[20] and consists of dissociation of the nuclear framework before Auger decay, which takes place in the atom.

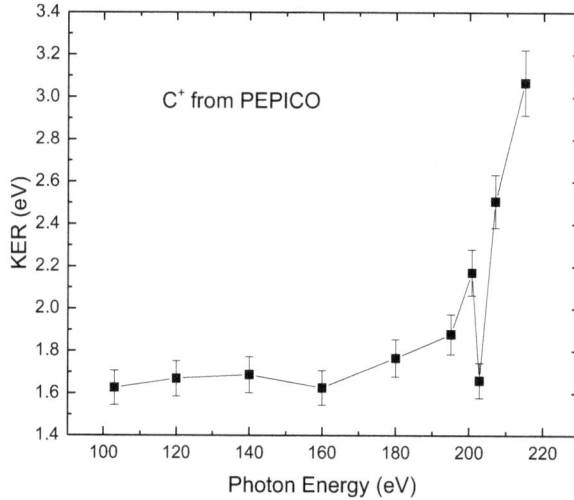

Fig. (8). Average center-of-mass KER (FWHM) of C^+ fragment from the CH_2Cl_2 molecule as a function of the incident photon wavelength.

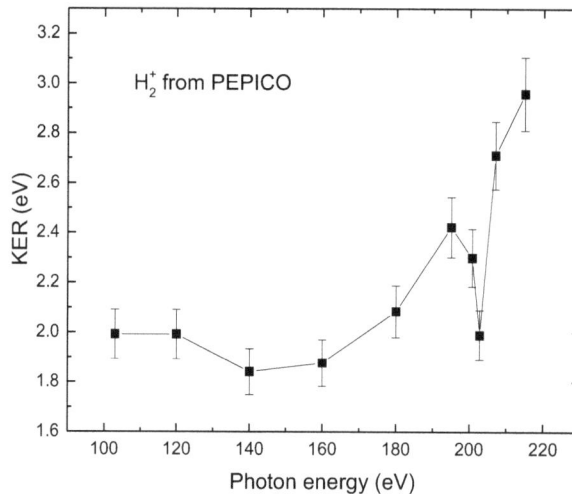

Fig. (9). Average center-of-mass KER (FWHM) of $H_2{}^+$ fragment from the CH_2Cl_2 molecule as a function of the incident photon wavelength.

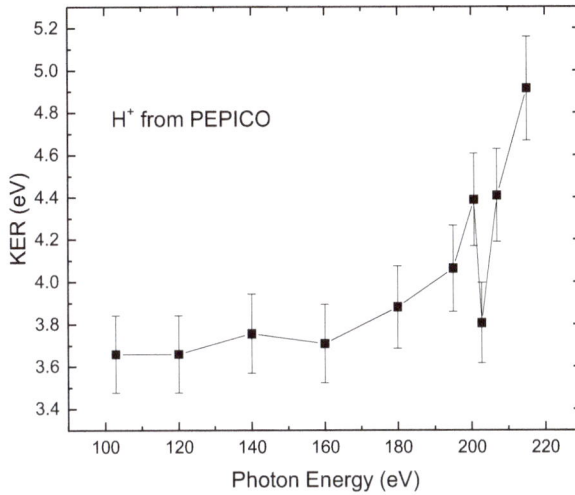

Fig. (10). Average center-of-mass KER (FWHM) of H⁻ fragment from the CH_2Cl_2 molecule as a function of the incident photon wavelength.

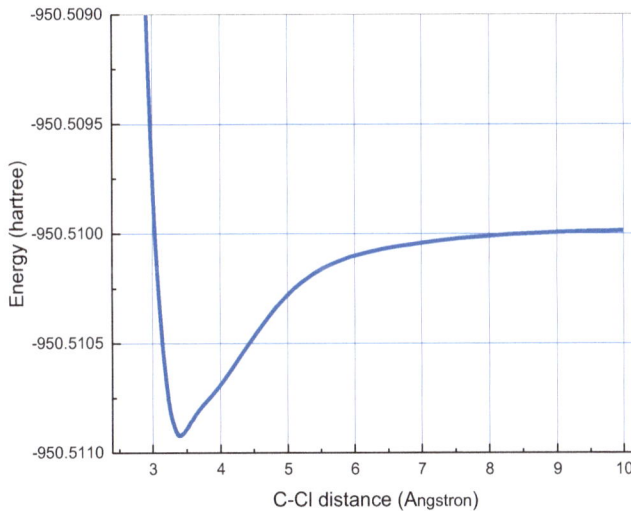

Fig. (11). Potential curve for inner-shell singlet Cl (L → ∞) ionized state. The total energy is calculated along the C-Cl distance. Each point represents the minimum energy, *i.e.*, that obtained by optimization of the other nuclear coordinates.

It is important to emphasize that for a given C-Cl distance all other molecular coordinates are optimized. Therefore, this is a minimum energy path to that state. For the state Cl (2p → ∞), that is inner-shell single ionization, a similar profile is obtained as also included in Fig. (**11**). A noticeable difference is the presence of a shallow minimum around 3.5 Å. In what concerns the discussion here, this difference is not relevant and an effective repulsive curve is also obtained. So, the process of ultra-fast dissociation $CH_2Cl + Cl^+$ is quite likely *via* the inner-shell curve.

Fig. (**12**) shows the potential curve for inner-shell singlet Cl (L $\rightarrow$ $10a_1^*$) excited state. The total energy is calculated along the C-Cl distance. Each point represents the minimum energy, *i.e.*, that obtained by optimization of the other nuclear coordinates

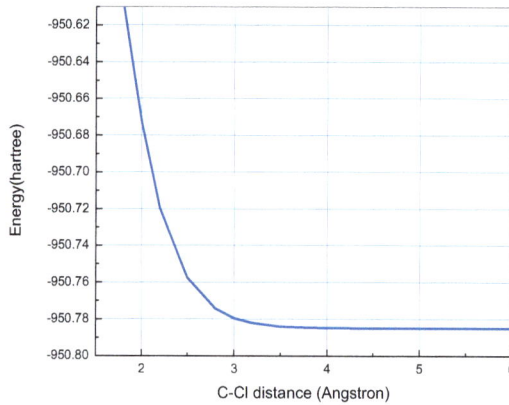

Fig. (12). Potential curve for inner-shell singlet Cl (L $\rightarrow$ $10a_1^*$) ionized state. The total energy is calculated along the C-Cl distance. Each point represents the minimum energy, *i.e.*, that obtained by optimization of the other nuclear coordinates.

Significant differences in the KER of the CH_2Cl_2 fragments can be also observed from Fig. (**13**) for the $L_{3/2} \rightarrow 10a_1^*$ and $L_{3/2} \rightarrow 4b_1^*$ transitions. This can be ascribed to the different symmetries of the $10a_1$ and $4b_1$ states. After excitation to the $10a_1$ or $4b_1$ states, the molecule decays *via* resonant Auger decay. As Auger processes are mainly driven by Coulombic interaction, their transition probability is given by

$$M_{nn'} = \left\langle n' \left| \frac{1}{r_{ij}} \right| n \right\rangle \tag{14}$$

where n and n' stand for initial and final states, respectively. It is worthy to emphasize that those states can have discrete and continuum parts. In the present case, the Coulombic repulsion operator transforms the totally symmetric irreducible representation of the point group C_{2v}. In order to determine if transitions probabilities are different from zero, one should examine the representation to which $|n\rangle$ and $|n'\rangle$ (or $\langle n'|$) belongs. The product of their irreducible representations should contain the totally symmetric representation. Since in C_{2v} there is no degenerate irreducible representation, the only possibility to achieve that condition is that the initial and final states belong to the same irreducible representation. Accordingly, if the initial core state is of the A_1 type, the final state has to be of the same type. The same reasoning applies if the initial state is of the B_1 type, *i.e.*, the final state has to belong to this representation. Therefore, it is not surprising to verify that the fragmentation is strongly dependent on the generated initial core state, since each state will follow specific Auger decay and consequently different fragmentation processes.

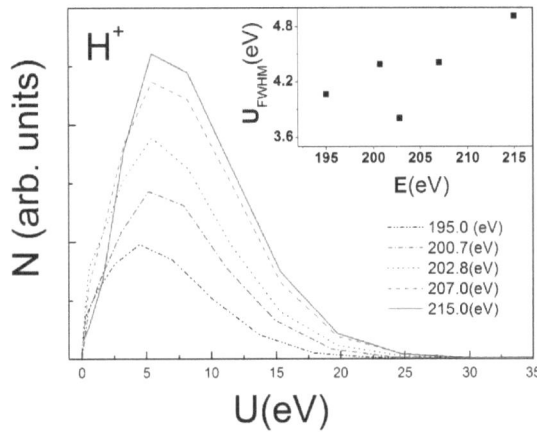

Fig. (13). Centre-of-mass KERD for the H^+ ion at selected photon energies around the Cl L edge. The corresponding average (FWHM) values are displayed as inset as a function of the photon wavelength E. The vertical scale was arbitrarily chosen for each distribution. Note that, despite this arbitrary choice regarding the vertical scale, this has not affected the value of U_{FWHM} that is plotted in the inset.

Kinetic energy distributions for the H^+ photo-dissociation fragments obtained at various photon energies are shown in Fig. (**13**). The inset in Fig. (**6**) displays the average KER obtained for the peak widths (FWHM) for comparison. The spectra show a sudden increase from 0 to around 5 eV, and drop more slowly after the maximum to 10% of their maximum intensity at 20 eV. The maximum kinetic energy value of the H^+ fragments reaches 25 eV. Since, the extraction field chosen for the spectrometer in this experiment is high enough to guarantee 100% collection efficiency for ions up to 30 eV, this abrupt cut-off cannot be attributed to a sharply decreasing collection efficiency. The high-energy components in the H^+ KERD might be associated to aborted double ionization events that are included in the PEPICO spectra. The spectrum obtained at 195 eV shows a peak at the 4.5 eV kinetic energy. Around the Cl L edge and above, the KERD peak is located at 5.5 eV. The average KER shows a different behavior, increasing from 4.0 eV below the edge, exhibiting a minimum of 3.8 eV at the $L_{3/2} \rightarrow 4b_1{}^*$ resonance (202.8 eV), and increasing again above the edge to 4.9 eV. The average

KER rising is in agreement with the assertion that an increase of the photon wavelength will populate a larger amount of the higher-excited molecular electronic states.

Fig. (**14**) shows the influence of the Cl (L → $10a_1^*$) and Cl (L → $4b_1^*$) transitions on the average KER of the C+, CH_2^+ and Cl^+ cations. The Cl (L → $4b_1^*$) transition at 202.8eV induces a full atomization break-up of the molecule and a sharp drop in the average KER of the C^+ cations is observed. On the other hand, the Cl (L → $10a_1^*$) transition at 200.7 eV leads to a less pronounced decrease of the average KER of the CH_2^+ cations. However, when the heavier Cl^+ cations are ejected, both transitions reduce the average KER with similar strengths, leading to a more shallow and wide KER decline.

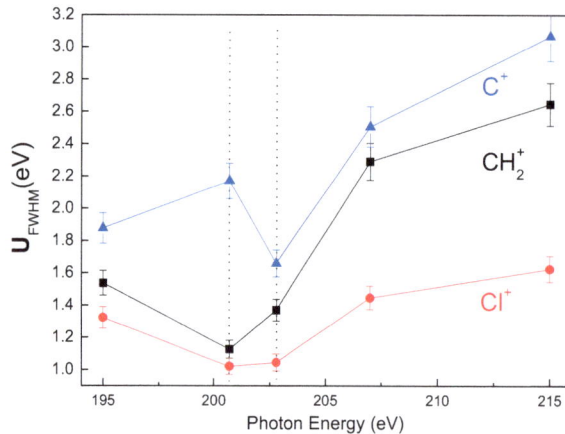

Fig. (14). Average KER for C^+, CH_2^+ and Cl^+ ions at selected photon energies around the Cl L edge. Dashed lines were included indicating the energies of the Cl (L → $10a_1^*$) and Cl (L → $4b_1^*$) transitions

The KERD of C^+, which corresponds to the fully atomization of the molecule, can be discerned in Fig. (**15**). In this case, a different tuning of the incident photon wavelength implies substantial KERD differences.

The structures in the KERD cannot be directly interpreted univocally due to the lack of theoretical potential energy curves of the parent ion. In a simple picture of instantaneous dissociation into several atomic fragments, the C^+ ions would hold very low kinetic energy due to its central position in the molecule. Notwithstanding, the measured translational KER of C^+ is not negligible. Then, it is evident that these non-negligible parts in the C^+ KER spectra (see Fig. **15**) are not coherent with the scenario of a concerted rupture of all bonds. One possibility is that the C^+ fragments are generated in a bent conformation, which, upon atomization, gives rise to C^+ fragment ions having significant KER. The vibrational bending modes of the molecule could likewise transfer a certain amount of KER. Finally, a sequential dissociation might also occur, where a sequential bond breaking takes place. The spectra display richer structures. For instance, bellow the Cl L edge, at 195 eV, the KERD presents bands at 0.21, 0.84, 2.8, 4.6 eV, and a tail on the high-energy side superimposed by a peak at 16 eV. This long tail implies the presence of a repulsive potential contributing to the formation of C^+ + neutrals in the Franck-Condon region. The rich structures and the large width of the KERDs show that there are several potential energy curves that are actually populated in the excitation process.

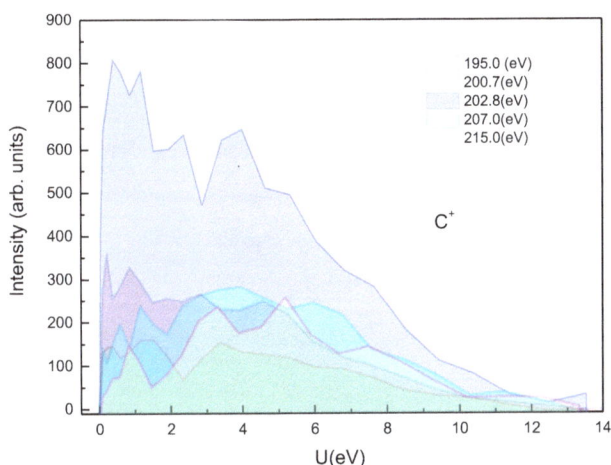

Fig. (15). KERD for the C^+ ion at selected photon energies around the Cl L edge.

Fig. (**16**) shows the KERD for the CH_2^+ fragment. The CH_2^+ KER spectra exhibit a small peak near zero eV and a maximum below 2 eV that moves to higher KER and becomes broader as the photon wavelength increases. Near the CH_2Cl_2 ionization energy and above, the spectra exhibit structures, which can be interpreted as due to the opening of several dissociative states of $CH_2Cl_2^+$ which evidence themselves in the KERD. The maximum KER lies between 12-14 eV, depending on the photon wavelength, as shown above in Fig. (**16**).

Fig. (**17**) displays the KERD for the Cl^+ fragment. The decreasing KERD for the heavier fragments is a consequence of the momentum conservation. At 195 eV, there are peaks centered at kinetic energies of 0.13, 0.51, and 0.97 eV, an intense peak centered at approximately 2.0 eV and a shoulder around 2.6 eV. The maximum translational energy is noted about 5 eV. At 200.7 eV, there are structures centered at 0.21, 0.68, 1.2, 2.1, 3.6, and 5.1 eV. At this energy, the average KER (see again Fig. **10**) decreases to about 0.18 eV. It can be seen that the KERDs of the Cl^+ fragment show a sharp increase in intensity from 0 to 0.2

eV at 195, 200.7, and 202.8 eV photon energies, while at 207 and 215 eV, this feature is less pronounced.

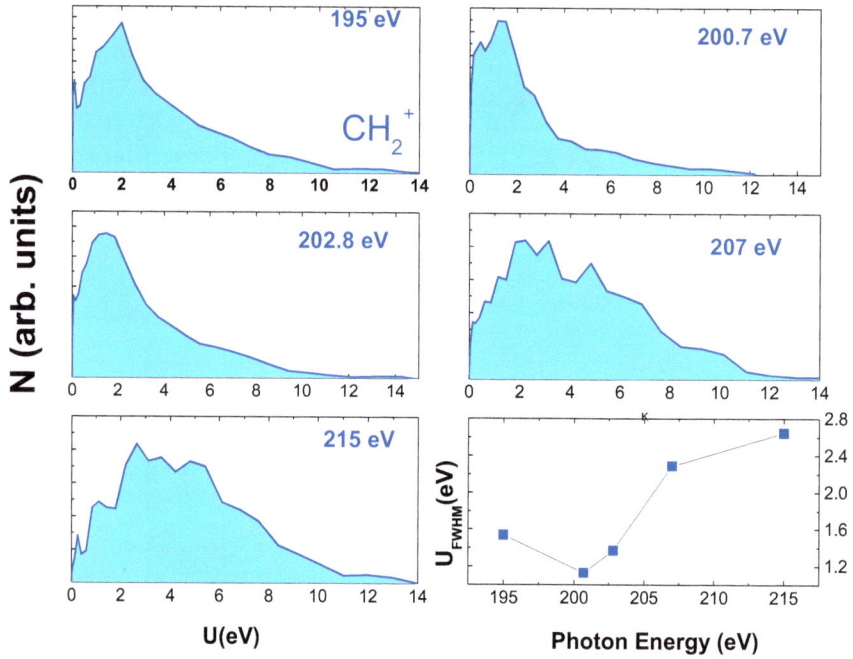

Fig. (16). KERD for the CH_2^+ ion at selected photon energies around the Cl L edge.

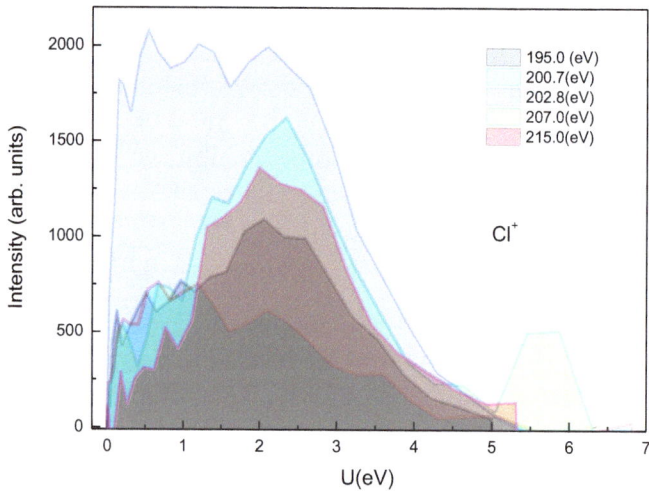

Fig. (17). KERD for the Cl^+ ion at selected photon energies around the Cl L edge.

IV.2. The Dication $CH_2Cl_2^{2+}$

Molecular dications are richer in electronic excited states in comparison to the isoelectronic atomic dications because of their lower symmetry [20, 21]. The fragmentation of the $CH_2Cl_2^{2+}$ gives rise, preferentially, to the formation of two singly charged fragments. Neglecting the inter-nuclear charge density effects [22], the inter-nuclear potential surface energy can be approximated by the Coulomb interaction between the fragment ions, of the charged fragments results in a larger KER. Besides its importance for electron correlation knowledge[10], double photoionization of molecules has the particularity that the Franck-Condon transition from the neutral gives rise to doubly charged moieties in a nuclear arrangement frequently exceedingly far from the nuclear equilibrium. Consequently, doubly charged polyatomic cations coming from double photoionization are commonly unstable moieties, which can dissociate into fragments with high kinetic energies, leading to striking results [23, 24]. The average KERs of several dissociation channels of the doubly charged $CH_2Cl_2^{2+}$ molecule are shown in Fig. (**11**) as a function of the photon wavelength. The corresponding average KER and KERDs were extracted after projecting the PELICO coincidences onto the T_1 and T_2 axis[10]. Due to momentum conservation, the lighter fragments carry larger KERs. It can be seen that the average KER for the $H^+ + Cl^+$ coincidence is narrower at both $L_{3/2}$ and $L_{1/2}$ resonances. In contrast, the peak widths for the $H^+ + CH^+$ coincidence are broader at those resonances. Below the Cl L resonance, the ratio of the momenta of H^+ and Cl^+ is $p_H^+/p_{Cl}^+ = -2.6$, which is compatible with a four-body secondary decay [25]:

$$CH_2Cl_2^{2+} \rightarrow CH_2Cl^+ + Cl^+ \ (U_1)$$

$$CH_2Cl^+ \rightarrow CH_2^+ + Cl \ (U_2)$$

$$CH_2^+ \rightarrow CH + H^+ \ (U_3)$$

In the case of the dissociation of the $CH_2Cl_2^+$, the predominant part of the KER is produced during charge separation. In that case, the KER can be approximately related to the nuclear distance between the fragments at the moment of the explosion, R, through the equation.

$$U = \frac{1}{4\pi\varepsilon_o} \cdot \frac{q_1 q_2}{R} \tag{15}$$

or

$$U(eV) = \frac{14.4 q_1 q_2}{R(\overset{o}{A})} \tag{16}$$

The neutral molecule (C_{2v}) geometries are [26, 27]: R(C-H) = 1.78 Å, R(C-Cl) = 1,77 Å, HCH = 112.0°, ClCCl=11.8°. Based on Eq. 16 and Fig. (**11**) one can estimates R ≈ 2.0 Å.

In the case of the $H^+ + C^+$ coincidence, the sum of the KER of both fragments lies between 8.3 and 9.7 eV, slightly larger than in the case of the former channel. This can be understood in terms of a smaller inter-nuclear distance between the H^+ and C^+ fragments, since carbon is the central atom. In the case of the $Cl^+ + Cl^+$ coincidence (not shown), both chlorine cations carry KER between 0.6 and 1.0 eV. They have the same KER (and the same momenta) only at the $L_{1/2}$ and $L_{3/2}$ resonances, indicating they are released in a secondary decay [25].

Figs. (**18-20** and **22**) show average KER (FWHM) of some coincidences after double ionization of the CH_2Cl_2 molecule as a function of the incident photon wavelength. Fig. (**21**) shows PEPIPICO spectrum of the $Cl^+ + CH_2Cl^+$ coincidence at 103 eV.

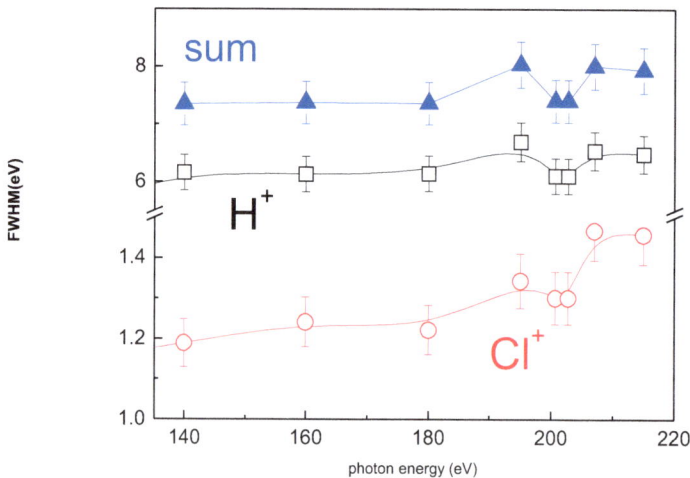

Fig. (18). Average KER (FWHM) of some coincidences after double ionization of the CH_2Cl_2 molecule as a function of the incident photon wavelength.

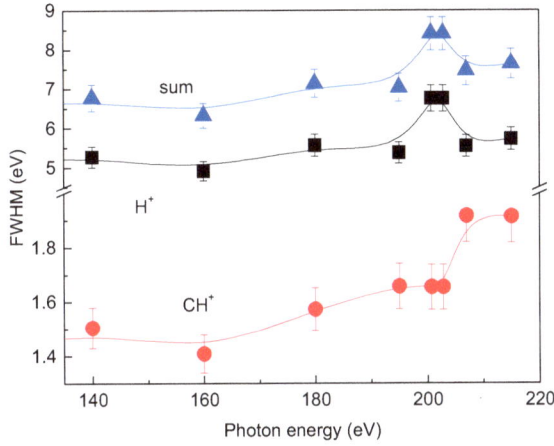

Fig. (19). Average KER (FWHM) of some coincidences after double ionization of the CH_2Cl_2 molecule as a function of the incident photon wavelength.

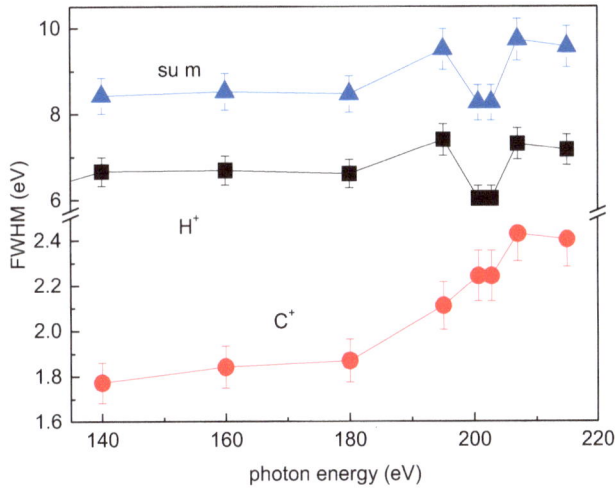

Fig. (20). Average KER (FWHM) of some coincidences after double ionization of the CH_2Cl_2 molecule as a function of the incident photon wavelength.

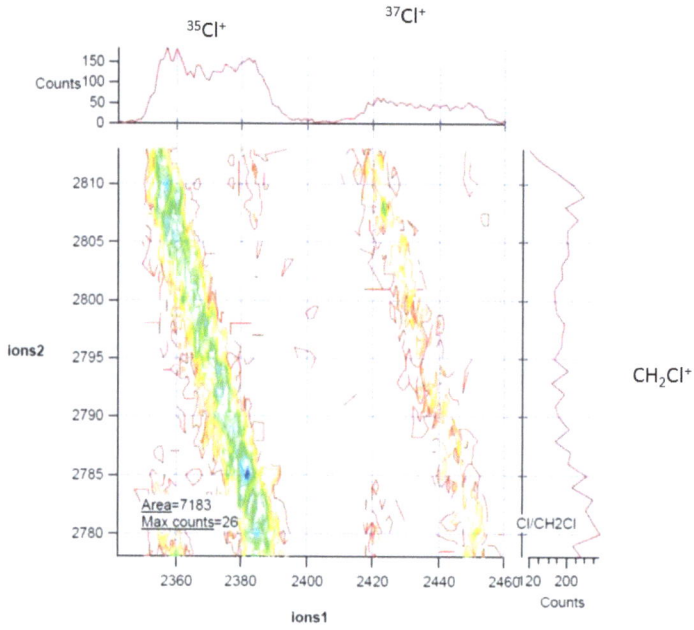

Fig. (21). PEPIPIICO spectrum of the $Cl^+ + CH_2Cl^+$ coincidence at 103 eV.

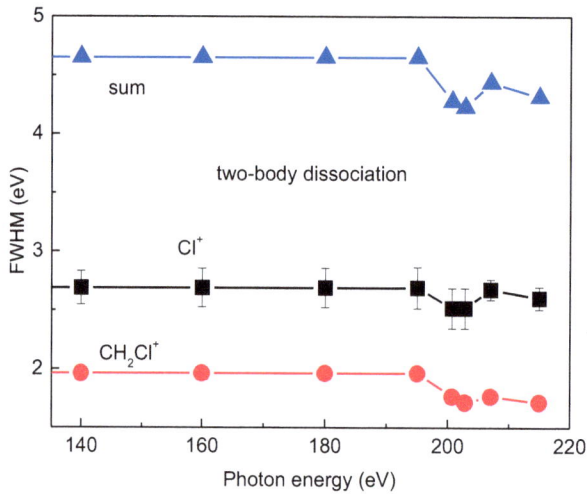

Fig. (22). Average KER (FWHM) of some coincidences after double ionization of the CH_2Cl_2 molecule as a function of the incident photon wavelength.

The $CH_2Cl^+ + Cl^+$ channel is usually considered as a Coulomb explosion process where, due to momentum conservation, the lighter fragment takes most of the energy. The KER of both fragments around the Cl L, present minimum KER, as shown in Fig. (**11**). The validity of the assumption that the process $CH_2Cl_2^{2+} \rightarrow CH_2Cl^+ + Cl^+$ is governed by a simple Coulomb explosion is evidenced by calculating the ground state potential curve, shown in Fig. (**12**). For a distance larger than 2.85 Å, the curve is repulsive, whilst for smaller distances there a potential well. This means that that the dication can dissociate by Coulomb explosion or can have a longer lifetime depending on the region it is formed, when analysing Fig. (**12**). As this state is formed from repulsive inner-shell state, which can lead to fast dissociation, it is plausible to admit that it will be predominantly formed in the repulsive region, *i.e.*, at distances larger than 2.85 Å. Thus, the assumption of pure Coulombic repulsion used above is justifiable. It is worth emphasizing the abrupt change of tendency of the potential curve at 2.85 Å. For larger distances the Cl^+ approaches CH_2Cl^+ in-plane with the two Chlorine atoms at maximum distance. At shorter distances, the most stable approach is out-of-plane, which favors bonding. This is also shown in Fig. (**12**).

Fig. (**24**) shows the KER distributions of the $H^+ + Cl^+$ dissociation channel arising from the fragmentation of the $CH_2Cl_2^{2+}$ precursor ion at selected photon energies around the Cl L edge. The measured H^+ ion KERD is very broad and described by a single peak located around 7.5 eV at hv= 195 eV, and rises as the photon wavelength increases, while, as expected, in single coincidence the peaks were observed around 5.5 eV (Fig. **24**). Both fragments present minimum average KER at the Cl L resonances. Comparing the average total energy values shown in Fig. (**11**) with the theoretical KER values obtained from Eq. 15 using typical internuclear distances (2.0 Å), a quite good agreement between experimental and theoretical values is achieved. The shape of the KERD profile of the H^+ moiety changes slightly as a function of the photon wavelength. The maximum KERD of the H^+ increases from 27 eV at hv=195 eV up to 30 eV at hv=215 eV. The KERD spectra of the Cl^+ fragment show several narrow peaks.

Fig. (**23**) shows the calculated potential energy curve for the ground state of dication $CH_2Cl_2^{2+}$. The total energy is calculated along the C-Cl distance. Each point represents the minimum energy, obtained by optimization of the other nuclear coordinates. Fig. (**23**) also shows, two selected configurations of the molecule at two different point of potential curve are shown: at 3.00 Å, the atoms are all in the same plane, while, at 2.85 Å, a Chorine atom is out-of-plane, in a bonding-favoring geometry.

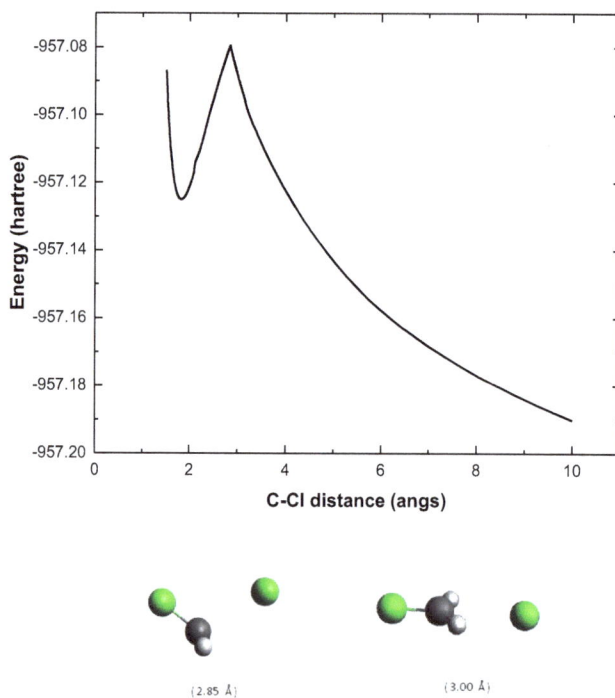

Fig. (23). On the left, potential energy curve for the ground state of dication $CH_2Cl_2^{2+}$. The total energy is calculated along the C-Cl distance. Each point represents the minimum energy, obtained by optimization of the other nuclear coordinates. On the right, two selected configurations of the molecule at two different point of potential curve are shown: at 3.00 Å, the atoms are all in the same plane, while, at 2.85 Å, a Chorine atom is out-of-plane, in a bonding-favoring geometry.

Fig. (**25**) displays the KERD of the $CH_2^+ + Cl^+$ fragmentation channel arising from the fragmentation of the parent molecule dication as precursor ion at selected photon energies around the Cl L edge. Both distributions present several structures. Again, the structures observed in the distributions are not presently amenable to unambiguous interpretation due to the unavailability of detailed potential energy curves of the doubly charged DCM molecule. Notwithstanding, as opposed to the other channels, the dissociation channel is formed *via* a non-Coulomb potential energy curve. A two-step model has been postulated in the literature [28] in which initial excitation of very highly charged molecular states are followed by rapid dissociation.

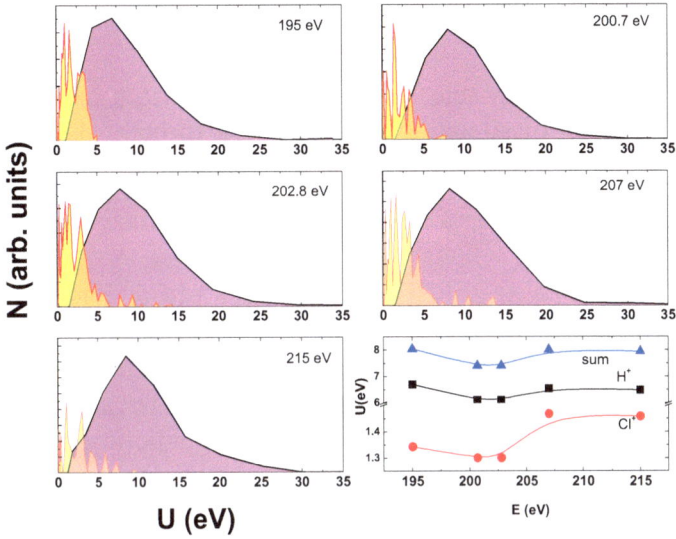

Fig. (24). KERD and average KER for the $H^+ + Cl^+$ coincidence ions at selected photon energies around the Cl L edge. Dash lines: H^+; Full lines: Cl^+.

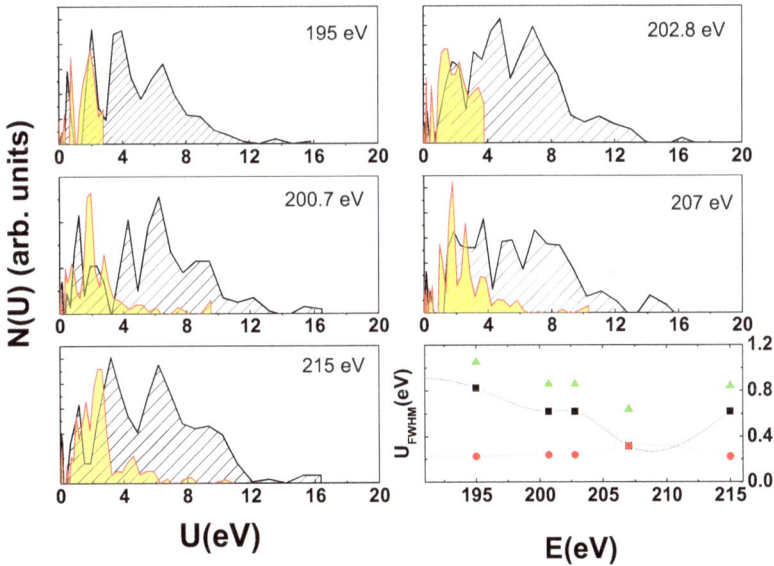

Fig. (25). KERD and average KER for the $CH_2^+ + Cl^+$ coincidence ions at selected photon energies around the Cl L edge. Dash lines: CH_2^+; full lines: Cl^+.

For more references on this subject, the reader is referred [28 - 35].

V. SUMMARY

In conclusion, the kinetic energy release is of major interest in molecular reaction dynamics. The KERD of dissociation products are directly associated to the derivative of the measured peak profile. Their determination therefore involves data of very high statistics and a relatively small instrumental width so as to provide accurate and reliable results.

When dissociation reactions of polyatomic ions take place, some of the excess internal energy of the ion is released as kinetic energy of the two fragments. The KER carries very valuable information concerning the structures of the species involved and concerning the energetics and dynamics of the reaction. In a two-body fragmentation of a singly charged ion the total kinetic energy release is imparted to the ionic as well as to the neutral fragment. The total is, however, calculable from the kinetic energy of the ionic fragment through simple momentum and energy conservation. On the other hand, in the many-body fragmentation process, where two or more neutral fragments are released, the total kinetic energy cannot be calculated. Thus, in this work, a systematic experimental and theoretical study of the excitation and fragmentation processes of the CH_2Cl_2 molecule was performed by the translational energy analysis of the ion fragments from the TOF peak shapes. Average KER and KERD are reported for the most abundant fragments in the PEPICO (Cl^+, H^+, and CH_2^+) and PE2PICO ($H^+ + Cl^+$, $H^+ + C^+$, $Cl^+ + Cl^+$, and $H^+ + CH^+$) spectra, the latter arising from incomplete Coulomb fragmentation. In the photon wavelength range selected in this investigation, the increase of the photon wavelength evolves in an accompanying change in the KERDs, either in intensity or kinetic energy shift, comprising very different processes. The H^+ fragment having the $CH_2Cl_2^+$ as a precursor is ascertained to have a unimodal kinetic energy dispersion in the range from 0 to 20 eV, with a maximum kinetic energy near 5.0 eV. Significant differences in the KER of the CH_2Cl_2 fragments for the $2p_{3/2} \rightarrow 10a_1^*$ and $2p_{3/2} \rightarrow 4b_1^*$ transitions were observed and ascribed to the different symmetries of the $10a_1$ and $4b_1$ states.

CONSENT FOR PUBLICATION

Not applicable.

CONFLICT OF INTEREST

The author (editor) declares no conflict of interest, financial or otherwise.

ACKNOWLEDGEMENTS

This work is supported in part by CNPq (Brazil).

REFERENCES

[1]   R. Singh, P. Bhatt, N. Yadav, and R. Shanker, "Ionic fragmentation of a CH4 molecule induced by 10-keV electrons: Kinetic-energy-release distributions and dissociation mechanisms", *Phys. Rev. A,* vol. 87, p. 062706, 2013.
[http://dx.doi.org/10.1103/PhysRevA.87.062706]

[2]   R. Singh, P. Bhatt, and N. Yadav, "Ionic fragmentation of the CO molecule by impact of 10-keV electrons: Kinetic-energy-release distributions", *Phys. Rev. A,* vol. 87, p. 022709, 2013.
[http://dx.doi.org/10.1103/PhysRevA.87.022709]

[3]   D. Mathur, "Structure and dynamics of molecules in high charge states", *Phys. Rep.,* vol. 391, p. 1, 2004.
[http://dx.doi.org/10.1016/j.physrep.2003.10.016]

[4]   A.C.F. Santos, W.S. Melo, M.M. Sant'Anna, G.M. Sigaud, and E.C. Montenegro, "Fragmentation and mean kinetic energy release of the nitrogen molecule", *Nuclear Instrum. Meth. B,* vol. 261, pp. 200-203, 2007.
[http://dx.doi.org/10.1016/j.nimb.2007.04.030]

[5]   R. Ravi, L. Philip, and T. Swaminathan, "Comparison of biological reactors (biofilter, biotrickling filter and modified RBC) for treating dichloromethane vapors", *J. Chem. Technol. Biotechnol.,* vol. 85, p. 634, 2010.
[http://dx.doi.org/10.1002/jctb.2344]

[6]   M.A. Hussain, R. Ford, and J. Hill, "Determination of fecal contamination indicator sterols in an Australian water supply system", *Environ. Monit. Assess.,* vol. 165, no. 1-4, pp. 147-157, 2010.
[http://dx.doi.org/10.1007/s10661-009-0934-5] [PMID: 19421885]

[7]   H. Lu, L. Zhu, and S. Chen, "Pollution level, phase distribution and health risk of polycyclic aromatic hydrocarbons in indoor air at public places of Hangzhou, China", *Environ. Pollut.,* vol. 152, no. 3, pp. 569-575, 2008.
[http://dx.doi.org/10.1016/j.envpol.2007.07.005] [PMID: 17698267]

[8]   K.F. Alcantara, W. Wolff, and A.H.A. Gomes, "Fragmentation of the CH2Cl2 molecule by proton impact and VUV photons", *J. Phys. At. Mol. Opt. Phys.,* vol. 44, p. 165205, 2011.
[http://dx.doi.org/10.1088/0953-4075/44/16/165205]

[9]   A.H.A. Gomes, W. Wolff, and N. Ferreira, "Deep core ionic photofragmentation of the CF2Cl2 molecule", *Int. J. Mass Spectrom.,* vol. 319, p. 1, 2012.
[http://dx.doi.org/10.1016/j.ijms.2012.03.004]

[10]   K.F. Alcantara, A.H.A. Gomes, and L. Sigaud, "Outer-shell double photoionization of CH2Cl2", *Chem. Phys.,* vol. 429, p. 1, 2014.
[http://dx.doi.org/10.1016/j.chemphys.2013.11.019]

[11]   A.C.F. Santos, M.G. Homem, D.P. Almeida, and J. Elec, "Production of highly charged Ne ions by synchrotron radiation", *Spect. Rel. Phen.,* vol. 184, p. 38, 2011.
[http://dx.doi.org/10.1016/j.elspec.2010.12.002]

[12]   A.B. Rocha, "Spin-orbit splitting for inner-shell 2p states", *J. Mol. Model.,* vol. 20, no. 8, p. 2355, 2014.
[http://dx.doi.org/10.1007/s00894-014-2355-9] [PMID: 25031078]

[13]   H.A. Bethe, and E.E. Salpeter, *Quantum Mechanics of the One and Two Electron Atoms.* Plenum: New York, 1977.
[http://dx.doi.org/10.1007/978-1-4613-4104-8]

[14]   T.R. Furlani, and H.F. King, "Theory of spin-orbit coupling. Application to singlet–triplet interaction in the trimethylene biradical", *J. Chem. Phys.,* vol. 82, p. 5577, 1985.
       [http://dx.doi.org/10.1063/1.448967]

[15]   D.G. Fedorov, and M.S. Gordon, "A study of the relative importance of one and two-electron contributions to spin–orbit coupling", *J. Chem. Phys.,* vol. 112, p. 5611, 2000.
       [http://dx.doi.org/10.1063/1.481136]

[16]   A. Berning, M. Schweizer, H-J. Werner, P.J. Knowles, and P. Palmieri, "Spin-orbit matrix elements for internally contracted multi-reference configuration interaction wave functions", *Mol. Phys.,* vol. 98, p. 1823, 2000.
       [http://dx.doi.org/10.1080/00268970009483386]

[17]   H.J. Werner, P.J. Knowles, G. Knizia, F.R. Manby, and M. Schütz, MOLPRO, v 2012.1, A Package of Ab Initio Programs; http://www.molpro.net

[18]   K.T. Lu, J.M. Chen, J.M. Lee, S.C. Haw, S.A. Chen, Y.C. Liang, and S.W. Chen, "State-selective enhanced production of positive ions and excited neutral fragments of gaseous CH2Cl2 following Cl 2p core-level photoexcitation", *Phys. Rev. A,* vol. 82, p. 033421, 2010.
       [http://dx.doi.org/10.1103/PhysRevA.82.033421]

[19]   K.F. Alcantara, A.H.A. Gomes, W. Wolff, L. Sigaud, and A.C.F. Santos, "Strong Electronic Selectivity in the Shallow Core Excitation of the CH2Cl2 Molecule", *J. Phys. Chem. A,* vol. 119, no. 33, pp. 8822-8831, 2015.
       [http://dx.doi.org/10.1021/acs.jpca.5b04402] [PMID: 26220163]

[20]   C. Miron, P. Morin, D. Céolin, L. Journel, and M. Simon, "Multipathway dissociation dynamics of core-excited methyl chloride probed by high resolution electron spectroscopy and Auger-electron-ion coincidences", *J. Chem. Phys.,* vol. 128, no. 15, p. 154314, 2008.
       [http://dx.doi.org/10.1063/1.2900645] [PMID: 18433216]

[21]   C.P. Safvan, and D. Mathur, "Dissociation of highly charged N2q+ (q >= 2) ions via non-Coulombic potential energy curves", *J. Phys. B,* vol. 27, p. 4073, 1994.
       [http://dx.doi.org/10.1088/0953-4075/27/17/028]

[22]   G. Dujardin, D. Winkoun, and S. Leach, "Double photoionization of methane", *Phys. Rev. A,* vol. 31, p. 3027, 1985.
       [http://dx.doi.org/10.1103/PhysRevA.31.3027]

[23]   R. Thissen, O. Witasse, O. Dutuit, C.S. Wedlund, G. Gronoff, and J. Lilensten, "Doubly-charged ions in the planetary ionospheres: a review", *Phys. Chem. Chem. Phys.,* vol. 13, no. 41, pp. 18264-18287, 2011.
       [http://dx.doi.org/10.1039/c1cp21957j] [PMID: 21931881]

[24]   L. Sigaud, N. Ferreira, and E.C. Montenegro, "Absolute cross sections for O2 dication production by electron impact", *J. Chem. Phys.,* vol. 139, p. 024302, 2013.
       [http://dx.doi.org/10.1063/1.4812779] [PMID: 23862938]

[25]   M. Simon, T. Lebrun, R. Martins, G.G.B. de Souza, I. Nenner, M. Lavolee, and P. Morin, "Multicoincidence mass spectrometry applied to hexamethyldisilane excited around the silicon 2p edge", *J. Phys. Chem.,* vol. 97, p. 5228, 1993.
       [http://dx.doi.org/10.1021/j100122a011]

[26]   R.P. Grant, F.M. Harris, and D.E. Parry, "An experimental and computational study of the double ionization of CH3Cl, CH2Cl2, and CHCl3 molecules to singlet and triplet electronic states of their dications", *Int. J. Mass Spectrom.,* vol. 192, p. 111, 1999.
       [http://dx.doi.org/10.1016/S1387-3806(99)00087-1]

[27]   R.J. Myers, and W.D. GWinn, "The Microwave Spectra, Structure, Dipole Moment, and Chlorine Nuclear Quadrupole Coupling Constants of Methylene Chloride", *J. Chem. Phys.,* vol. 20, p. 1420, 1952.

[28]   W.T. Hill III, J. Zhu, D.L. Hatten, Y. Cui, J. Goldhar, and S. Yang, "Role of non-Coulombic potential curves in intense field dissociative ionization of diatomic molecules", *Phys. Rev. Lett.,* vol. 69, no. 18, pp. 2646-2649, 1992.
[http://dx.doi.org/10.1103/PhysRevLett.69.2646] [PMID: 10046548]

[29]   W.J. Griffiths, and F.M. Harris, "Experimental determination of the double ionization energies of the chloromethanes CH3Cl, CH2Cl2 and CHCl3", *Rapid Commun. Mass Spectrom.,* vol. 22, pp. 91-94, 1988.
[http://dx.doi.org/10.1002/rcm.1290020507]

[30]   J.H.D. Eland, Dynamics of Double Photoionization in Molecules and Atoms.*Advances in Chemical Physics.,* S.A. Rice, Ed., vol. Vol. 141. John Wiley & Sons, Inc.: Hoboken, NJ, USA, 2009.
[http://dx.doi.org/10.1002/9780470431917.ch3]

[31]   O. Witasse, O. Dutuit, and J. Lilensten, "Correction to "Prediction of a CO_2^{2+} layer in the atmosphere of Mars"", *Geophys. Res. Lett.,* vol. 30, p. 1360, 2003.
[http://dx.doi.org/10.1029/2003GL017007]

[32]   J. Lilensten, C.S. Wedlund, and M. Barthélémy, "Dications and thermal ions in planetary atmospheric escape", *Icarus,* vol. 222, pp. 169-187, 2013.
[http://dx.doi.org/10.1016/j.icarus.2012.09.034]

[33]   S. Falcinelli, M. Rosi, and P. Candori, "Kinetic Energy Release in molecular dications fragmentation after VUV and EUV ionization and escape from planetary atmospheres", *Planet. Space Sci.,* vol. 99, pp. 149-157, 2014.
[http://dx.doi.org/10.1016/j.pss.2014.04.020]

[34]   S. Falcinelli, F. Pirani, and M. Alagia, "The escape of O^+ ions from the atmosphere: An explanation of the observed ion density profiles on Mars", *Chem. Phys. Lett.,* vol. 666, pp. 1-6, 2016.
[http://dx.doi.org/10.1016/j.cplett.2016.09.003]

[35]   S. Falcinelli, "The Escape of O^+ and CO^+ Ions from Mars and Titan Atmospheres by Coulomb Explosion of CO_2^{2+} Molecular Dications", *Acta Phys. Pol. A,* vol. 131, no. 1, 2017.
[http://dx.doi.org/10.12693/APhysPolA.131.112]

CHAPTER 3

Selective Excision of Biomolecules in Electron Transfer Experiments: Current Developments and Achievements

Paulo Limão-Vieira[1,*], Filipe Ferreira da Silva[1] and **Gustavo García[2]**

[1] *Laboratório de Colisões Atómicas e Moleculares, CEFITEC, Departamento de Física, Faculdade de Ciências e Tecnologia, Universidade NOVA de Lisboa, 2829-516 Caparica, Portugal*

[2] *Instituto de Fisica Fundamental, Consejo Superior de Investigaciones Científicas, Serrano 113-bis, 28006 Madrid, Spain*

Abstract: Here we revisit electron transfer processes yielding negative ion formation in gas-phase collisions of fast neutral potassium atoms (electron donor) and biomolecular target molecules (electron acceptor) in a crossed molecular-beam arrangement. The negative ions formed in the interaction region are time-of-flight (TOF) mass analysed as a function of the collision energy. Selective site and bond excision in the unimolecular decomposition of the transient negative show clear dependence on the collision energy.

Keywords: Anions, Attachment, Atom-molecule collisions, DNA/RNA subunits, DNA-strand breaks, Electron transfer, Electrons, Fragmentation, Gas-phase, Metastable decay, TOF mass spectrometry, Radiation damage.

INTRODUCTION

The alterations induced by high energy radiation in biological systems, in particular within living cells, DNA/RNA and other key biological targets, are now known to be essentially produced by the secondary species generated along the radiation track and their consecutive reactions within irradiated cells [1]. These species, *e.g.* excited atoms and molecules, radicals, ions, secondary electrons, may give rise to genotoxic, mutagenic, and several potentially lethal DNA lesions [2], such as base and sugar changes, single strand breaks, base release, and cluster lesions, including a combination of two single modifications, double strand breaks and cross-links. From the intermediate species created within nanoscopic volumes

* **Corresponding author Paulo Limão-Vieira:** Laboratório de Colisões Atómicas e Moleculares, CEFITEC, Departamento de Física, Faculdade de Ciências e Tecnologia, Universidade NOVA de Lisboa, 2829-516 Caparica, Portugal; Tel: (+351) 21 294 78 59; E-mail: plimaovieira@fct.unl.pt

Antônio Carlos Fontes dos Santos (Ed.)

along the ionisation tracks ($\sim 4 \times 10^4$ per 1 MeV incident particle), secondary electrons are known to be the most abundant [3 - 5].

The vast majority of these secondary electrons, after some thermalisation (through successive inelastic interactions) in the medium in question are created with energies below 30 eV [5, 6], producing large quantities of highly reactive radicals, cations, and anions [7] or can alternatively, in its final thermalisation process prehydrate in the medium itself to induce further chemical changes [8] with higher quantum yield than hydrated electrons and OH radicals [9]. These species are found to be more efficient generating degradation than the primary radiation; *i.e.* they are more reactive. As far as radiation damage is concerned, it appears extremely relevant to assess a comprehensive description of the interaction processes and their implications, by properly addressing the number of dissociative events, type of radicals generated and accurate energy and angular distribution functions for those species interacting with the medium [10]. However, a key and fundamental aspect of plasmid DNA irradiation by low-energy electrons for determining quantum yields for cell damage lies on the assessment of DNA strand breaks, with a description of the underlying molecular mechanisms [1]. Yet, electron interactions in particular sites within DNA showed a resonant behaviour rather than a monotonic enhancement above threshold as it happens for ionisation processes. Though, dissociative electron attachment (DEA) has been identified as a crucial damaging mechanism to a site in DNA, with subsequently decay of the transient negative ion (TNI) formed $(M^*)^-$, leading to dissociation or alternatively competing with autodetachment, leaving the molecule in an electronically excited state (M^*) [7, 11].

However, many elementary collisional processes are not due to direct impact but rather depend upon electron transfer. Studying chemical reactions for biomolecular systems is important to understand radiation induced damage at the molecular level with the outermost need to develop more efficient radiation therapies. Electron induced chemistry is also prevalent in many natural and industrial processes in a wide variety of media, including the formation of organic molecules within ice mantles on dusty grains in the interstellar medium [12]; the control of fluorocarbon plasmas used produce silicon chips [13] and the chemical modification of absorbates using electron patterning [14] and scanning tunnel microscopy [15], just to mention a few.

The former activities performed within the scope of electron transfer have led the authors to make unprecedented achievements, in the comprehension of several underlying molecular mechanisms yielding fragmentation of selected molecular targets upon electron transfer to biomolecules in atom-molecule collision experiments. Such achievements include novel electron transfer induced fragmentation patterns of thymine and uracil [16], with the most striking difference from previous DEA results being the enhanced yield of anions stemming from bond breaks in the ring. Of relevance, in terms of controlling and inducing selectivity of chemical reactions in molecular collisions, it was shown recently for the first time that at room temperature and with random molecular orientation, site (N1-H *vs.* N3-H) and bond (C-H *vs.* C-N) selective dissociation in DNA/RNA bases can be achieved by tuning the proper collision energy [17]. Later experiments on 1- and 3-methylthymine/methyluracil, have shown that NCO⁻ branching ratios as a function of the collision energy are reminiscent of extraordinary site- and bond-selectivity in the reactions yielding its formation [18]. These findings allowed to establish a new collision induced dissociation mechanism for DNA damage, which may be described at a basic molecular level. This model also provides a coherent explanation of the observed correlation between electron transfer to biomolecules and their carcinogenicity, and may be used to suggest new compounds to be adopted in radiation therapy as treatment enhancing sensitizers.

Radiosensitization properties of halouracils (*e.g.* 5-XU, X = F, Cl, Br, I) have been known for several decades with irradiation of cells in which some DNA thymines have been replaced by halogenated uracils, increasing the frequency of both single and double strand breaks [19]. The electron transfer model provides an explanation for such effects with the introduction of a strongly electrophilic atom into the DNA (*e.g.* F, Cl), leading to an enhancement in the collision induced dissociation probability and thence an increased probability for DNA destruction in cells containing such compounds [20]. In order to obtain the fragmentation patterns in the DNA/RNA sugar unit analogue, and assess the major significance of the decomposition mechanisms, our other contributions include investigations into tetrahydrofuran (THF) [21] and D-ribose [22]. Here we have observed ring breaking as the main decomposition channel, in contrast to results from DEA experiments [23]. Special emphasis was also given to the dissociation mechanisms lending support to the breaking of the N-glycosidic bond, as an initial step in the fragmentation of the TNI in uridine [24]. These studies have shown that electron transfer processes are much more effective than DEA in producing loss of integrity within DNA/RNA units. As far as the small aminoacids are concerned, neutral OH loss in glycine was observed [25] whereas in small aliphatic aminoacids (alanine and valine), the differences observed are due to the higher number of degrees of freedom of the side chain, in the case of valine, that can be linked with the formation of lighter fragments when the fragmentation process proceeds through a statistical dissociation [26]. In DEA studies to several (bio)molecular targets, the dominant fragmentation channels result from very low-energy resonances (often as low as $\square$0 eV) consisting of vibrational Feshbach resonances [27]. This can be rationalized by the fact that, in DEA, accessing high-energy resonances (such as NCO⁻ formation in uracil/thymine [23]) will mostly result in autodetachment, rather than in fragmentation. However, in atom (K) - molecule (ABC) collisions, there is evidence that autodetachment is significantly suppressed, due to the Coulomb interaction in the collision complex (K⁺ABC⁻) enhancing fragment formation. Such is also the case for D-ribose [22].

In sum, we report negative ion time-of-flight (TOF) mass spectra from collisions of neutral potassium atoms with several molecular targets. The negative ions result from a transfer of an electron from the neutral potassium to the target molecule (ABC) as simply represented by

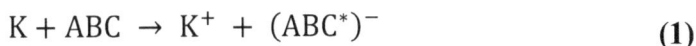

$$K + ABC \rightarrow K^+ + (ABC^*)^-$$ (1)

where K stems for the potassium atom and ABC for the polyatomic target molecule.

FUNDAMENTAL ASPECTS OF ELECTRON TRANSFER PROCESSES

In atom-molecule collisions, transfer processes happen when electrons follow adiabatically the nuclear motion in the vicinity of the crossing of the stationary nonperturbed states [28], *i.e.* the covalent and the ionic diabatic states. For simplicity let us consider a diatomic molecule, although for polyatomics hyperdimensional surfaces must be similarly considered. The ionic surface lies above the covalent surface, the endoergicity at large atom-molecule distances being $\Delta E = IE(K) - EA(ABC)$; with $IE(K)$ the ionisation energy of the potassium atom and $EA(ABC)$ the molecule's electron affinity. Due to the Coulombic interaction in the collision complex, there is a crossing point for which both stationary nonadiabatic potential energy surfaces have the same value [28]. For the lowest covalent and ionic states, the crossing radius of the diabatic potential energy surfaces can, roughly speaking, be obtained through [29]:

$$R_c = \frac{e^2}{\Delta E} = \frac{14.42}{\Delta E} \ [\text{Å}]$$ (2)

where ΔE represents the endoergicity. In cases where the collision time is approximately identical to the vibrational period, the crossing radius is shifted outwards during the collision, increasing the probability of a diabatic transition at the second crossing and thereby enhancing ion-pair formation. Conversely, if the collision time is smaller than the vibrational period, the molecular geometry can be assumed to be frozen [28].

During the collision process and near that crossing (R_c), there can be a perturbation of the stationary states induced by the projectile or target nuclear motion leading to an adiabatic coupling. This leads, after the collision path, to the formation of a positive ion K^+ and a molecular TNI, allowing access to parent molecular states which are not accessible in free-electron attachment experiments [29, 30]. In particular states with a positive electron affinity can be formed, and the role of vibrational excitation of the parent neutral molecule can be studied by the collision dynamics [30]. Even if the free negative molecular ion unstable in the gas phase, in the atom-molecule collision complex it can be stabilized [17]. The TNI lifetime will depend upon both the collision time (several tenths of fs) and the autodetachment time (tens of fs). If the lifetime of the TNI is longer than the fragmentation time, energy can be distributed over the available vibrational degrees of freedom and so change the fragmentation pathways. If the collision time is shorter than the dissociation time, collision induced dissociation is likely to take place and produce fragment ions with finite kinetic energies.

EXPERIMENTAL SET-UP

The experimental setup used for electron transfer studies to biomolecules in the gas phase is shown in Fig. (**1**). The setup is of a crossed atom-molecule beam arrangement consisting of a potassium source, an oven, and a linear time-of-flight (TOF) mass analyser [31]. The components are housed in two high-vacuum chambers at a base pressure of 10^{-5} Pa. A neutral potassium beam at an energy resolution of ~0.5 eV (FWHM), generated from a charge exchange chamber, intersects orthogonally with an effusive molecular beam consisting of the target molecules.

The neutral potassium beam is generated in the following way. Atomic K^+ ions, obtained from a commercial potassium ionic source (operating at ~1100 K), are accelerated (K^+_{hyp}) through a chamber containing potassium vapour (K^o_{th}) where they are resonantly charge exchanged to form a beam of fast neutral K atoms (K^o_{hyp}), schematically represented as:

$$K^+_{hyp} + K^o_{th} \rightarrow K^o_{hyp} + K^+_{th} \qquad (3)$$

The energy of the resultant K neutral beam is established by the initial acceleration of the ions. After charge exchange, the ions that have not been neutralized are removed by electrostatic fields, the resulting neutral K molecular beam is now comprised of a hyperthermal beam (K^o_{hyp}) and an effusive thermal energy beam (K^+_{th}). Since the electron transfer process is endoergic, the thermal beam does not contribute to the formation of anions. The hyperthermal alkali

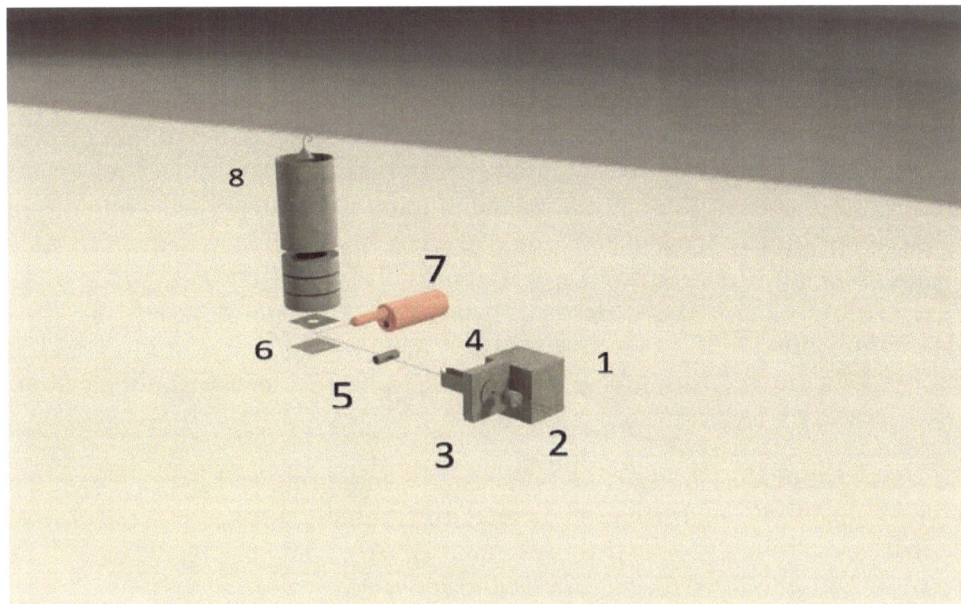

Fig. (1). Schematic view of the experimental setup used for electron transfer experiments, (1) neutral potassium oven, (2) potassium ion source, (3) charge exchange chamber, (4) electrostatic deflecting plates, (5) hyperthermal neutral potassium beam detector (of Langmuir-Taylor type), (6) ion deflecting plate, (7) neutral biomolecule oven, and (8) TOF mass spectrometer.

beam now enters a high-vacuum chamber where it is monitored by an iridium surface ionisation detector of the Langmuir-Taylor type. This detector samples the beam intensity, does not interfere with the beam passing to the collision region and it operates in a temperature regime that only allows detection of the fast beam. The biomolecular target beam is produced in a hot gas cell and admitted to vacuum by an effusive source through a 1 mm diameter orifice, where it crosses the neutral hyperthermal potassium beam. Oven temperatures are monitored by a platinum resistance thermometer (Pt100) in order to guarantee that the density of intact molecules is high enough to yield a reasonable negative-ion signal. The negative ions produced in the collision region are extracted by a 250 V/cm pulsed electrostatic field towards the entrance of a 1.4 m linear TOF, where they are analysed and detected in a single-pulse counting mode. The spectra collected at each collision energy, showing the recorded anionic signals, are obtained by subtracting the background signal from the sample signal. Mass calibration is obtained through the well-known anionic species formed in potassium collisions with nitromethane molecules [32].

EXPERIMENTAL RESULTS AND DISCUSSION

The main focus of electron transfer investigations in potassium-molecule collision experiments, has been devoted to single DNA/RNA nucleobases and its derivatives as well as other relevant biomolecular targets such as single chain aminoacids due to their role as building blocks of proteins. Such research interests follow a prompt need of the international community to comprehensively describe the underlying molecular mechanisms triggered by secondary species produced along the ionising radiation tracks, with particular attention to the role of low-energy electrons with key biological targets [1]. The comparison of the fragmentation profiles by electron transfer in atom-molecule studies and free electron molecule collisions will accordingly provide us to establish a more detailed model of electron induced damage in DNA [33].

In the next sections we present and discuss the most relevant achievements and aspects of negative ion formation in electron transfer processes, starting with nucleobases, followed by radiosensitizers, DNA/ RNA sugar units and their derivatives and ending with simple aminoacids.

Nucleobases

The nucleobases explored in electron transfer experiments by atom-molecule collisions, comprise the single aromatic ring pyrimidines (uracil and tymine), their methylated derivatives (*e.g.* 1- and 3-methylthymine/methyluracil), halogenated pyrimidines (5-XU, X = F, Cl), and recently with preliminary key studies on adenine as part of purines [34]. In the unimolecular decomposition of such pyrimidine-like molecular targets, we note four major interconnected and relevant mechanisms: a) dehydrogenated parent anion formation; b) ring breaking enhancement with the highest yield assigned to NCO^- formation; c) dehydrogenated parent anion as a precursor in the formation of the fragments that require bond cleavages in the ring, namely NCO^-; d) site- and bond-selective decomposition of the TNI by proper tuning the collision energy.

Electron transfer to isolated nucleobases does not lead to a stable molecular anion within mass spectrometric timescales [16 - 18, 34]. Instead these TNIs undergo autodetachment or dissociation into a negative fragment ion and one or more neutral species. Neutral potassium collisions (30–100 eV) with gas phase thymine (T), $C_5H_6N_2O_2$, and uracil (U), $C_4H_4N_2O_2$, reveal that the two most abundant product anions are assigned to NCO^- and the dehydrogenated anion $(U-H)^-$ / $(T-H)^-$. In Fig. (**2**), we restrict ourselves to the TOF mass spectrum at 30 eV collision energy, although we have performed measurements at 70 and 100 eV in the lab frame [16]. Increasing energy in potassium-thymine/uracil interactions will produce vibrationally excited states of the low-lying anion state, which may lie in the continuum of the neutral molecule giving rise to autodetachment or even result in further fragmentation. This is only possible if a stabilizing agent is close enough to the TNI to allow efficient intramolecular electron transfer. The anion mass spectra for thymine and uracil, roughly speaking, show similar fragmentation patterns. The similarities are not surprising, given that the molecules' configuration (and geometry) of the neutral states of these species is quite similar, as are the wavefunctions of their LUMOs [35].

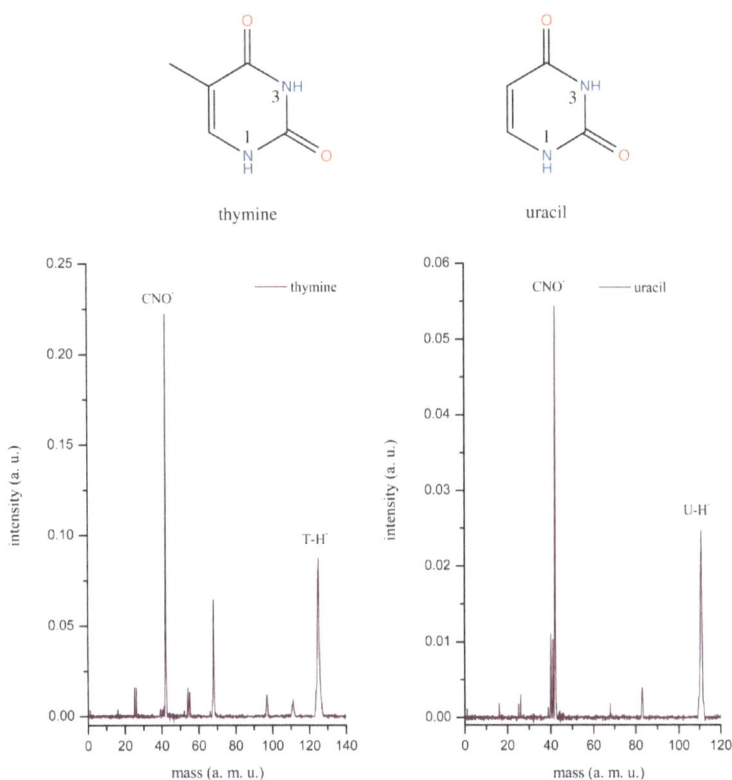

thymine   uracil

Fig. (2). Negative ion TOF mass spectra for potassium collisions with gas phase thymine and uracil at 30 eV (lab frame).

The striking difference from the present measurements to previous DEA studies to these molecules corresponds mainly to the relative intensities of the fragments that result from ring-breaking. While the formation of (T–H)⁻ / (U–H)⁻, and H⁻ can be rationalized in terms of simple bond cleavages, the other fragments result from more complex reactions, typically involving multiple bond cleavages in the ring and intramolecular energy redistribution among the available degrees of freedom. Previous studies [36, 37] support the efficient autodetachment of electrons from the orbitals that are responsible for the breaking of ring bonds.

Moreover, the electron transfer results may be rationalized in terms of a longer interaction time with the potassium cation stabilising the molecular anion with respect to autodetachment, and hence allowing more efficient electron transfer to the ring-breaking σ^* orbitals [38, 39]. Such is only possible if the nuclear wavepacket survives long enough for the π^* state to diabatically couple with an σ^* state leading to dissociation. As so, this is in assertion that the dehydrogenated parent anion is indeed a precursor in the formation of the fragments that require bond cleavages in the ring, namely NCO^-.

H⁻ Site- and Bond-selective Formation

The ion yields (relative intensity as a function of the collision energy) of H⁻ from 1-meT, 3-meU and D⁻ yields from thymine deuterated at the C positions (thymine-d4), are shown in Fig. (**3**) at three different collision energies.

Charge transfer deposited on gas-phase thymine, uracil, partly deuterated thymine, methylated thymine at the N1 (1-meT) and methylated uracil at the N3 (3-meU) positions by electron transfer mechanism, induces the loss of hydrogen which exclusively takes place from the N positions. The bond selectivity may alike be formed site selectively by appropriate adjustment of the collision energy. While at 5.3 eV collision energy results in the loss of hydrogen from N1 in 3-methyl-uracil, the reaction can be suppressed from N3 by tuning the collision energy to 7.6 eV as is in 1-methylthymine. Moreover, D⁻ formation from thymine deuterated in the C positions is suppressed at 7.4 eV showing that H⁻ formation in 3-methyl-uracil proceeds only through the N1 position. Here, energy and charge transfer are completely inactivated when the N1-H bond is replaced by N1-CH3.

In DEA experiments to thymine, the electron energy resonance profiles show that the minimum energy required to break a N1–H, N3–H, C6-H, and CH2–H bond, lies between 4 - 5 eV [40], so bond- and site-selectivity [41] to gas phase methylated and deuterated pyrimidines yielding H⁻ formation does not result from any particular energy constraint [42]. Since energy constraints cannot explain site-selectivity, the electronic structure of the associated TNIs accessed by electrons of different energies (either shape or core excited resonances) has been suggested as the main effect responsible for such an achievement. Taking the adiabatic electron affinities of 1-meT, 3-meU, and thymine-d4, as (0.025 ± 0.010) eV [43], (0.035 ± 0.010) eV [43] and (0.069 ± 0.007) eV [44], the values for R_c from equation (**2**) are found for the three molecular targets of the order of 3.3 Å. The corresponding total cross sections for ion-pair formation will be $\sim \pi R_c^2$, which is much larger than the corresponding gas kinetic cross sections.

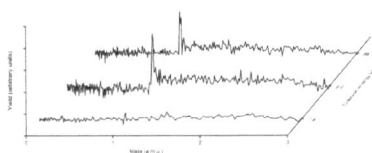

a)

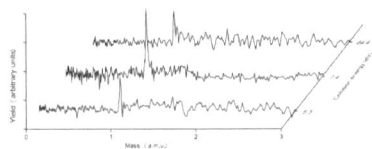

b)

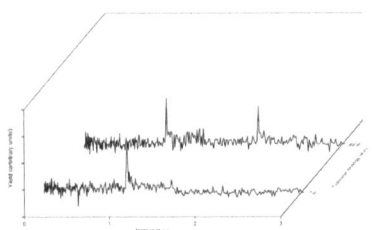

c)

Fig. (3). Negative ion TOF mass spectra yielding: (a) H⁻ formation from thymine methylated at the N1 position (1-meT) at 7.6, 9.0 and 66.1 eV; (b) H⁻ formation from uracil methylated at the N3 position (3-meU) at 5.3, 7.4, and 64.4 eV; (c) H⁻ and D⁻ formation from partly deuterated thymine (thymine-d4) at 7.4 and 64.9 eV.

NCO⁻ Site- and Bond-selective Formation

In potassium collisions with 1- and 3-methylthymine and 1- and 3-methyluracil, the dominant fragment above 30 eV collision energy is the NCO⁻ ion and the branching ratios as a function of the collision energy show evidence of extraordinary site-selectivity in the reactions yielding its formation (see Fig. **4**).

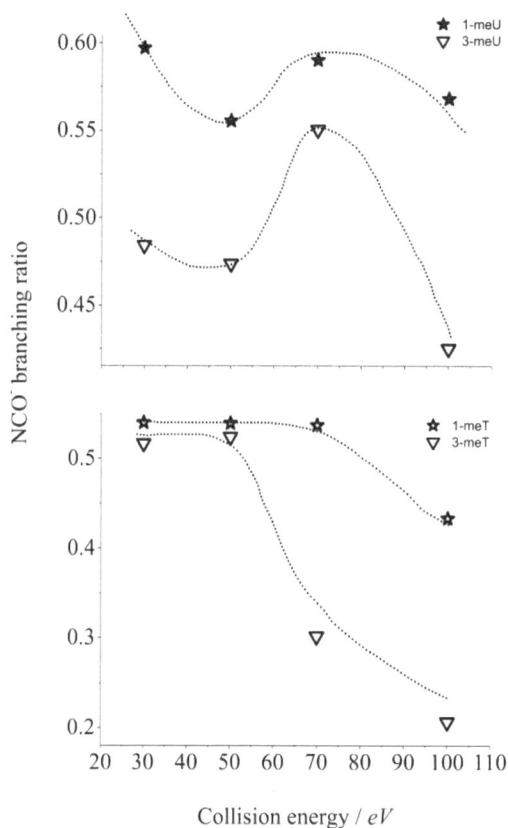

Fig. (4). NCO⁻ branching ratios in collisions of potassium atoms with N-site methylated uracil and thymine molecules. The lines are just to guide the eye.

Fig. **(5)** shows the reaction scheme of the three different possible reaction pathways leading to NCO⁻ formation in thymine upon electron capture calculated at the B2PLYP/ma-TZVP level of theory [18].

In the case of collision energies below 50 eV, (interaction times longer than 40 fs) the potassium cation (K^+) produced after electron transfer to the target molecule may be present sufficiently long in order to stabilize the TNI through Coulomb interaction within the collision complex. This ensures that pathways (d) and (e) are open together with pathway (c). This rationale can be interpreted in terms of the spatial orientation of the molecular orbitals. The incoming electron can be trapped in the C5 methyl group position π^* orbital, with time sufficient to be transferred to σ* N-CH3 from the N3 position methyl group. This assumption seems reasonable since a similar mechanism has been described in the context to the loss of a methyl group in potassium collisions with 1-meT and 3-meU [45]. Though, in the unimolecular decomposition such coupling will allow accessing the three pathways (c), (d), and (e) in Fig. (**5**), which in turn means similar NCO⁻ branching ratio for energies up to 50 eV. For higher collision energies (70 and 100 eV), the K⁺ stabilizing effect will not be effective enough because of its reduced transit time, and this coupling may no longer hold. In this energy range, the relative difference between the NCO⁻ branching ratios from 1-meT and 3-meT is, thus, increased to about 0.2 in favour of NCO⁻ production from 1-meT. This effect is mostly noticeable for energies above 50 eV and these results show an unprecedented site- and bond-selectivity regarding the concerted fragmentation mechanism yielding NCO⁻ formation in DNA pyrimidine bases upon electron transfer.

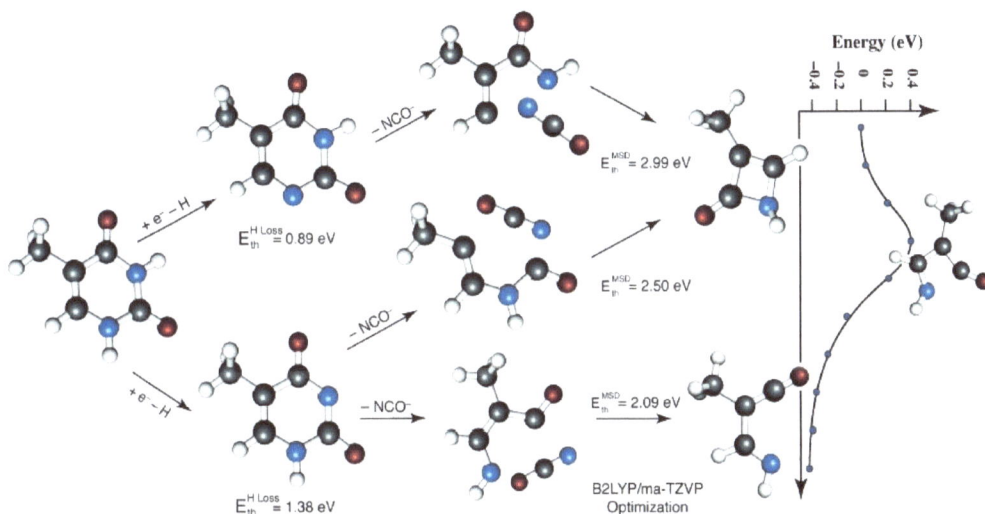

Fig. (5). Reaction scheme showing the three different possible pathways leading to NCO⁻ formation in thymine upon electron capture, along with the calculated threshold energies for individual reaction steps [18]. (a) H loss from N1; (b) H loss from N3; (c), (d), and (e) pathways leading to NCO⁻ formation.

In sum, these findings point to a new achievement in controlling chemical reactions that may have particular relevance for the investigation of early molecular processes in the nascent stages of DNA damage by secondary electrons, especially those related to strand breaks. Such selectivity, as far as bond and site are concerned, were reported for the first time by electron transfer induced dissociation experiments in alkali-molecule collisions [17, 18].

Radiosensitizers

The development of concomitant chemical and radiation therapy techniques revealed to be an important tool to improve the efficiency of cancer treatments. The simultaneous use of chemotherapy with radiotherapy can enable lower doses to be used when compared with radiotherapy used as a single treatment. This process involves a radiosensitizer that may enhance viral DNA/RNA damage. Halogenated pyrimidines (halouracils) have attracted considerable interest from the medical community as radiosensitizers, since they can be sensitive to the ionising radiation and increase the rate of cell death upon irradiation.

Electron transfer to 5-chlorouracil (5-ClU) and 5-fluorouracil (5-FU) yielding anion formation has been investigated in the collision energy region of 30–100 eV (not shown here) [20]. The rich fragmentation patterns suggests that electron transfer in collisions with electronegative neutrals may cause efficient damage to RNA. The main ring fragment anion in all mass spectra was assigned to NCO⁻ while the production of X⁻ (X = F, Cl) was identified as a strong decomposition of the halouracil temporary negative ions. Cl⁻ was the most intense fragment anion in the 5-ClU measurements, whereas NCO⁻ production dominated in the 5-FU data. The reaction leading to chloride formation was identified to be endothermic, whereas the equivalent reaction producing fluoride from 5-FU exothermic. The NCO⁻/ X⁻ (X = F, Cl) branching ratios, showed that Cl⁻ formation is independent of the collision time, whereas for shorter collision times, F⁻ yield is enhanced. With the exception of low-energy (30 eV) potassium collisions with 5-ClU, the molecules can be treated as rigid polar systems with respect to X–U stretching coordinate during the collision since the period of the vibrational mode is longer than the interaction time. Comparisons of this data with previous DEA results have shown that the presence of potassium plays an important role in the sort and number of fragment anions formed [20].

From the biological point of view, these results may have very interesting relevance, since decomposition of these targets in potassium collisions yields appreciable ring breaking (NCO⁻), which in turn may compromise the integrity of such molecular systems such as in viral RNA. Moreover the appreciable formation of the halogen anions is also relevant making these compounds well attuned for radiosensitization.

DNA/RNA Sugar Units and Their Derivatives

Studying the processes that occur in the context of electron transfer in atom-molecule collisions can be a stepping stone in our understanding of some of the molecular mechanisms that occur in non-gas-phase environments. In particular, studying the role of the sugar unit is critical, as it is now well-established that one of the main sources of possible damage to DNA/RNA stems from changes in the D-ribose (DR) unit (Fig. **6**).

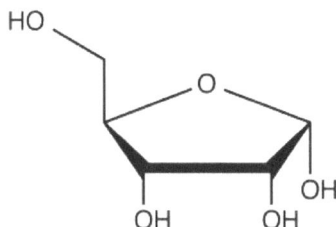

Fig. (6). Molecular structure of D-Ribose (DR), $C_5H_{10}O_5$.

The electron transfer mechanism is only possible at particular potassium-D-ribose molecule distances, the crossing radius, R_c, with a rough estimate of ~3.2 Å. Negative ion formation from collisions of neutral potassium atoms with D-ribose $(C_5H_{10}O_5)$, the monosaccharide pentose ring in the DNA/RNA structure, reveals that OH^- is the main fragment detected in the collision range 50–100 eV accounting on average for 50% of the total anion yield. Prominence was also given to the rich fragmentation pattern observed with special attention to O^- formation. These results are in sharp contrast to previous DEA experiments [46]. Noteworthy is the fact that neither the parent nor its dehydrogenated negative ions are reported in potassium-DR collisions. This lends support to the fragile nature of the sugar ring within the context of DNA. However, the relative yields of several fragments are significantly different when compared to DEA, in particular for OH^-, which in potassium-DR collisions is the dominant ion detected. The enhancement in the formation of this fragment is proposed to be due to the ability of potassium cation to suppress and/or even delay efficiently autodetachment, an effect that has been increasingly observed as pervasive in the context of atom-molecule collision studies [16, 20, 25, 47].

It is known that the sugar unit in DNA has a furanosic form and studies with such sugars, namely, THF (tetrahydrofuran, C_4H_8O), are essential to ascertain how important this characteristic is in the discussion of DNA/RNA sugar substitutes. Though, we have engendered a concerted study on THF and for a detailed and comprehensive description see [21] and references therein. Briefly, the data strongly highlights the major differences in the fragmentation dynamics between DEA and electron transfer experiments (Table **1**). It is worth noting that the DEA and electron transfer fragmentation patterns for THF are the most dissimilar for all of the molecular targets studied so far.

Table 1. Assignment of the fragment anions produced in collisions of potassium atoms with THF (C_4H_8O, 72 a.m.u.) in an energy range 20−100 eV against DEA experiments [48].

Mass peak (a.m.u.)	Assignment	
	This study (electron transfer)	DEA [48]
1	H^-	H^-
13	CH^-	–
14	CH_2^-	CH_2^-
16	O^-	O^-
17	OH^-	OH^-
24	C_2^-	–
25	C_2H^-	–
41	$C_3H_5^-$	$C_3H_5^-$
43	$C_2H_3O^-$	$C_2H_3O^-$
71	–	$C_4H_7O^-$
72	–	$C_4H_8O^-$

Actually, some DEA studies report the formation of the parent anion and its dehydrogenated analogue [48] both absent in the context of electron transfer studies. This point highlights a tendency to observe an enhancement of ring breaking in electron transfer experiments, as opposed to DEA, where the dominant fragmentation normally does not proceed through loss of ring integrity.

The electron transfer data indicates that collision-frame threshold energies for the production of particular fragment anions are broadly consistent with the onsets of corresponding DEA resonances. This suggests that the initially accessed states in DEA and electron transfer are the same. However, it is clear that these states can decay into different fragments, presumably due to the presence of the potassium cation in the vicinity of the TNI. In fact, it is suggested that the fragmentation of THF is sequential whereupon a chain ester anion is initially produced by C−O bond break. This is followed by a subsequent cleavage along the different bonds in the alkyl chain. Depending on the site of bond break, different fragments are formed. For example, a bond break of the remaining C−O bond will most likely lead to O^- production, whereas cleavages`1 of the various C−C bonds will entail the formation of all other fragments (with the exception of OH^- and H^-). We then tentatively suggested that fragmentation will most consist of an intramolecular electron transfer from the initial TNI (intermediate state) into different highly antibonding valence states along the C−C bonds. Owing to the similar ring structure, THF appears to be a good sugar unit surrogate in the context of electron-transfer processes, mainly as far as the ring breaking dynamics are concerned. This is further supported by comparing the DEA resonance profiles of THF in the gas phase [49, 50] with the profiles for single- and double-strand breaks of DNA [1]. Finally, the fragmentation patterns of D-ribose [22] have shown the importance of the hydroxyl groups as the main anionic species formed in electron-transfer experiments. As such, while a possible surrogate for the sugar unit, the use of THF will not be able to provide information on the relevance of the hydroxyl groups in the actual DNA/RNA sugar unit.

Single Chain Aminoacids

Electron transfer in potassium collisions with alanine ($C_3H_7NO_2$) and valine ($C_5H_{11}NO_2$) molecules was investigated at 15 and 100 eV. The fragmentation patterns obtained in the unimolecular decomposition through TOF mass spectrometry were compared for both amino acids as a function of the collision energy (Table **2**). A close comparison of the ionic yields allows describing the role of the side chain in the fragmentation pattern (Fig. (**7**)).

Table 2. Assignment of the fragment anions produced in collisions of potassium atoms with alanine (A) and valine (V) at 15 and 100 eV.

Mass peak (a.m.u)	Assignment	Collision energy (eV)	
		15 eV	*100 eV*
1	H⁻	A / V	A / V
12	C⁻	A / V	A / V
13	CH⁻	A / V	A / V
14	CH_2^-	–	A / V
16	O⁻ / NH_2^-	A / V	A / V
17	OH⁻	A / V	A / V
24	C_2^-	A / V	A / V
25	C_2H^-	A / V	A / V
26	$C_2H_2^-$ / CN⁻	A / V	A / V
41	CHCO⁻	A / V	A / V
45	COOH⁻ / CHOO⁻	–	A
48	Metastable decay	A / V	V
71	$C_3H_3O_2^-$ / $C_4H_9N^-$	A / V	V
88	$CH_3CH(NH_2)COO^-$ / [A−H]⁻	A	A
116	$C_3H_5O_2^-$ / [V−H]⁻	V	V

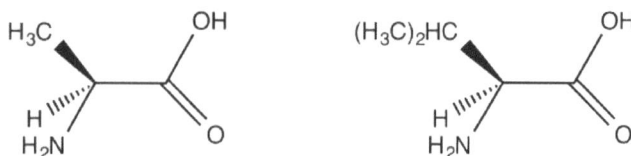

Fig. (7). Chemical structure of alanine ($C_3H_7NO_2$) and valine ($C_5H_{11}NO_2$).

In the case of alanine, the dehydrogenated parent anion ($[A-H]^-$) is one of the most intense fragments, whereas for valine ($[V-H]^-$) it corresponds to a minor fragment. One possible explanation for this difference is the higher number of degrees of freedom of the side chain in the case of valine that can be linked with the formation of lighter fragments when the fragmentation process proceeds through a statistical dissociation. Fragment anions 16 and 26 a.m.u. were associated to isobaric fragments, and their formation considered in terms of the available energy. Fragment 48 a.m.u. was reported for the first time and suggested to be attributed to metastable decay from a heavier anion [26].

Electron transfer in potassium collisions with the simplest amino acid glycine was investigated in the energy range from 20 to 100 eV [25]. The main characteristic in the TOF mass spectra was the relative decrease of the dehydrogenated parent anion yield with increasing collision energy. For low collision energies, we attributed the formation of fragment anions to delayed autodetachment of the TNI as a result of the Coulombic stabilisation of the (K^+ Glycine$^-$) complex. For higher collision energies, the fragmentation pattern has been rationalized in terms of the increasing available energy, as well as to metastable decay of heavier fragments (especially the dehydrogenated parent anion) in the unimolecular decomposition upon electron transfer. These have been assigned to masses 12.5, 18 and 23 a.m.u. The assignment of isobaric fragments 15, 16 and 26 a.m.u. has been discussed in the light of the available energy in the potassium glycine centre-of-mass system. In such collisions we have reported novel fragments anions at 12, 13, 24 and 25 a.m.u. which have been assigned to C^-, CH^-, C_2^- and C_2H^-, respectively [25].

CONCLUSIONS AND FUTURE WORK

The set of novel electron transfer experiments on potassium-biomolecule collisions have contributed to the scientific knowledge on: (a) site- and bond-selective excision processes in pyrimidines and their derivatives, through H abstraction and yielding NCO$^-$ formation; (b) the DNA/RNA loss of integrity enhancement with high yields of fragment anions in contrast to the dehydrogenated parent anion formation, the latter a prevalent mechanism in DEA experiments; (c) the possible stabilizing effect of K^+ while in the vicinity of the TNI, dictating the fragmentation patterns with clear discrepancies from the DEA environment; (d) the role of increasing side-chains in aminoacids enhancing small fragment anion yields in detriment to the dehydrogenated parent anion formation; (e) observation of other fragment anions not reported before, and to which we have suggested a metastable decay process from heavier precursor anions. These accomplishments, although extremely relevant within the general context of the collision dynamics, led to knowledge being gained on the underlying mechanisms in negative ion formation, and the role of electron transfer to biomolecular targets. Nonetheless there remain several unanswered questions, and processes that have been proposed to explain various observations still need to be properly addressed and probed. Although we have previously obtained negative ion yields and proposed the decomposition mechanisms of several biomolecules (concisely described above), the home-made linear TOF imposed instrumental limitations as to the mass resolution (~150) which meant it was unable to fully resolve and in some cases assign and explore particular dissociation channels. Of relevance are those within 1 a.m.u. difference, not only for lower masses but also for heavier species where the TOF mass peaks natural broadening increases. Most of the earlier achievements would benefit from a proper dedicated investigation of the underlying electronic states yielding a particular fragment anion and neutral species. The key points that need to be fully explored and are on-going are:(i) the recent increase in mass resolution will open up the possibility to further explore molecular anions within 1 a.m.u. difference and increase the total mass range detection; (ii) this will allow us to explore anion states yielding neutral radicals such as $H^\bullet$ and $H^\bullet + H^\bullet$ or H_2, as well as $O^\bullet$ and $OH^\bullet$ formation, and infer on their relevance within the biological environment; (iii) isobaric fragment detection that has been unresolved up to now; (iv) clarifying whether some anions are the result of metastable decay, which will lead us to assess the nature of the precursor state; (v) the role of the strong Coulomb interaction in the collision complex delaying autodetachment; (vi) exploring the stabilizing effect of K^+ in the vicinity of the TNI, thus allowing clarification of the intramolecular electron transfer process and the nature of the excited electronic states. The latter is still an unsolved case within the international community, in particular whether an σ^* or a π^* state (or even both) are prevalent in some electron transfer mechanisms within most the DNA/RNA pyrimidine units. In order to tackle these challenges, aiming to get full knowledge of these processes, here are combining two experimental approaches to investigate potassium-biomolecule collision experiments: (1) installing a high-resolution mass (> 3000) reflectron TOF spectrometer; (2) resolve metastable decay ambiguities through reflectron TOF; (3) installing a forward direction K^+ energy loss analyser detection system. Regarding the former, it will allow a reflectron operation mode in high mass resolution but also a linear mode to explore kinetic energy release distributions of a fragment anion profile. Such information will be valuable to explore particular resonances and their role in the dissociation process. As far as the K^+ energy loss system is concerned, a set of profiles as a function of the scattering angle are being planned for future investigations. In this angular description we will have a full picture of the collision dynamics, where we can explore the role of the covalent and ionic scattering processes involved in potassium-biomolecule collisions. These will give unprecedented results in the context of electron transfer to biomolecules. Currently, we rely on theoretical support from well-established colleagues within the community, with whom we have already probed some novel mechanisms in thymine and uracil [35]. These partnerships are certainly to be strengthen in the years to come.

CONSENT FOR PUBLICATION

Not applicable.

CONFLICT OF INTEREST

The author (editor) declares no conflict of interest, financial or otherwise.

ACKNOWLEDGMENTS

PLV and FFS acknowledge the Portuguese Foundation for Science and Technology (FCT-MEC) through sabbatical and post-doctoral grants, SFRH/BSAB/105792/2014 and SFRH/BPD/68979/2010, as well as the research grants PTDC/FIS-ATO/1832/2012 and UID/FIS/00068/2013. FFS acknowledges FCT-MEC for IF-FCT IF/00380/2014.

PLV together with GG acknowledge the Spanish-Portuguese joint collaboration through Project HP2006-0042. We also acknowledge the support of the Spanish Ministerio de Economia y Competitivad under Project No. FIS 2012-31230 and the European Union COST Action CM1401, Our Astro-Chemical History. PLV acknowledges his visiting professor positions at Sophia University, Tokyo, Japan, Flinders University, Adelaide, South Australia and The Open University, United Kingdom; GG acknowledges his professor position at the University of Wollongong, New South Wales, Australia. PLV and GG are also in debt to the Australian National University, Canberra, Australia.

REFERENCES

[1]   B. Boudaïffa, P. Cloutier, D. Hunting, M.A. Huels, and L. Sanche, "Resonant formation of DNA strand breaks by low-energy (3 to 20 eV) electrons", *Science,* vol. 287, no. 5458, pp. 1658-1660, 2000. [http://dx.doi.org/10.1126/science.287.5458.1658] [PMID: 10698742]

[2]   C. von Sonntag, *Free-Radical-Induced DNA Damage and Its Repair.* Springer: New York, 2005.

[3]   V. Cobut, Y. Frongillo, J.P. Patau, T. Goulet, M-J. Fraser, and J-P. Jay-Gerin, "Monte Carlo simulation of fast electron and proton tracks in liquid water - I. Physical and physicochemical aspects",

Radiat. Phys. Chem., vol. 51, pp. 229-243, 1998.

[4]   I. Abril, R. Garcia-Molina, C.D. Denton, I. Kyriakou, and D. Emfietzoglou, "Energy loss of hydrogen- and helium-ion beams in DNA: calculations based on a realistic energy-loss function of the target", *Radiat. Res.,* vol. 175, no. 2, pp. 247-255, 2011.
[http://dx.doi.org/10.1667/RR2142.1] [PMID: 21268719]

[5]   J.A. LaVerne, and S.M. Pimblott, "Electron energy-loss distributions in solid, dry DNA", *Radiat. Res.,* vol. 141, no. 2, pp. 208-215, 1995.
[http://dx.doi.org/10.2307/3579049] [PMID: 7838960]

[6]   S.M. Pimblott, and J.A. LaVerne, "Production of low-energy electrons by ionizing radiation", *Radiat. Res.,* vol. 76, pp. 1244-1247, 2007.

[7]   I. Baccarelli, I. Bald, F.A. Gianturco, E. Illenberger, and J. Kopyra, "Electron-induced damage of DNA and its components: Experiments and theoretical models", *Phys. Rep.,* vol. 508, pp. 1-44, 2011.
[http://dx.doi.org/10.1016/j.physrep.2011.06.004]

[8]   L. Turi, and P.J. Rossky, "Theoretical studies of spectroscopy and dynamics of hydrated electrons", *Chem. Rev.,* vol. 112, no. 11, pp. 5641-5674, 2012.
[http://dx.doi.org/10.1021/cr300144z] [PMID: 22954423]

[9]   C.R. Wang, J. Nguyen, and Q.B. Lu, "Bond breaks of nucleotides by dissociative electron transfer of nonequilibrium prehydrated electrons: a new molecular mechanism for reductive DNA damage", *J. Am. Chem. Soc.,* vol. 131, no. 32, pp. 11320-11322, 2009.
[http://dx.doi.org/10.1021/ja902675g] [PMID: 19634911]

[10] G.G. Gómez-Tejedor, and M.C. Fuss, *Radiation Damage in Biomolecular Systems.* Springer: New York, 2012.
[http://dx.doi.org/10.1007/978-94-007-2564-5]

[11] F. Martin, P. D. Burrow, Z. Cai, P. Coutier, D. Hunting, and L. Sanche, "DNA strand breaks induced by 0-4 eV electrons: The role of shape resonances", *Phys. Rev. Lett.,* vol. 93, no. 068101, pp. 1-4, 2004.

[12] M. Orzol, I. Martin, J. Kocisek, I. Dabkowska, J. Langer, and E. Illenberger, "Bond and site selectivity in dissociative electron attachment to gas phase and condensed phase ethanol and trifluoroethanol", *Phys. Chem. Chem. Phys.,* vol. 9, no. 26, pp. 3424-3431, 2007.
[http://dx.doi.org/10.1039/b701543g] [PMID: 17664966]

[13] D. Duflot, M. Hoshino, P. Limão-Vieira, A. Suga, H. Kato, and H. Tanaka, "BF3 valence and Rydberg states as probed by electron energy loss spectroscopy and ab initio calculations", *J. Phys. Chem. A,* vol. 118, no. 46, pp. 10955-10966, 2014.
[http://dx.doi.org/10.1021/jp509375y] [PMID: 25338148]

[14] S. Engmann, M. Stano, S. Matejčík, and O. Ingólfsson, "Gas phase low energy electron induced decomposition of the focused electron beam induced deposition (FEBID) precursor trimethyl (methylcyclopentadienyl) platinum(IV) (MeCpPtMe3)", *Phys. Chem. Chem. Phys.,* vol. 14, no. 42, pp. 14611-14618, 2012.
[http://dx.doi.org/10.1039/c2cp42637d] [PMID: 23032785]

[15] P.A. Sloan, and R.E. Palmer, "Two-electron dissociation of single molecules by atomic manipulation at room temperature", *Nature,* vol. 434, no. 7031, pp. 367-371, 2005.
[http://dx.doi.org/10.1038/nature03385] [PMID: 15772657]

[16] D. Almeida, R. Antunes, G. Martins, S. Eden, F. Ferreira da Silva, Y. Nunes, G. García, and P. Limão-Vieira, "Electron transfer-induced fragmentation of thymine and uracil in atom-molecule collisions", *Phys. Chem. Chem. Phys.,* vol. 13, no. 34, pp. 15657-15665, 2011.
[http://dx.doi.org/10.1039/c1cp21340g] [PMID: 21796297]

[17] D. Almeida, F. Ferreira da Silva, G. García, and P. Limão-Vieira, "Selective bond cleavage in potassium collisions with pyrimidine bases of DNA", *Phys. Rev. Lett.,* vol. 110, no. 023201, pp. 1-5,

2013.
[http://dx.doi.org/10.1103/PhysRevLett.110.023201]

[18]   F. Ferreira da Silva, C. Matias, D. Almeida, G. García, O. Ingólfsson, H.D. Flosadóttir, B. Ómarsson, S. Ptasinska, B. Puschnigg, P. Scheier, P. Limão-Vieira, and S. Denifl, "NCO(-), a key fragment upon dissociative electron attachment and electron transfer to pyrimidine bases: site selectivity for a slow decay process", *J. Am. Soc. Mass Spectrom.,* vol. 24, no. 11, pp. 1787-1797, 2013.
[http://dx.doi.org/10.1007/s13361-013-0715-9] [PMID: 24043519]

[19]   H. Abdoul-Carime, P. Limão-Vieira, I. Petrushko, N.J. Mason, S. Gohlke, and E. Illenberger, "5-Bromouridine sensitization by slow electrons", *Chem. Phys. Lett.,* vol. 393, pp. 442-447, 2004.
[http://dx.doi.org/10.1016/j.cplett.2004.06.081]

[20]   F. Ferreira da Silva, D. Almeida, R. Antunes, G. Martins, Y. Nunes, S. Eden, G. García, and P. Limão-Vieira, "Electron transfer processes in potassium collisions with 5-fluorouracil and 5-chlorouracil", *Phys. Chem. Chem. Phys.,* vol. 13, no. 48, pp. 21621-21629, 2011.
[http://dx.doi.org/10.1039/c1cp22644d] [PMID: 22071464]

[21]   D. Almeida, F. Ferreira da Silva, S. Eden, G. García, and P. Limão-Vieira, "New fragmentation pathways in K-THF collisions as studied by electron-transfer experiments: negative ion formation", *J. Phys. Chem. A,* vol. 118, no. 4, pp. 690-696, 2014.
[http://dx.doi.org/10.1021/jp407997w] [PMID: 24400742]

[22]   D. Almeida, F. Ferreira da Silva, G. García, and P. Limão-Vieira, "Dynamic of negative ions in potassium-D-Ribose collisions", *J. Chem. Phys.,* vol. 139, no. 214305, pp. 1-6, 2013.

[23]   S. Ptasińska, S. Denifl, P. Scheier, and T.D. Märk, "Inelastic electron interaction (attachment/ionization) with deoxyribose", *J. Chem. Phys.,* vol. 120, no. 18, pp. 8505-8511, 2004.
[http://dx.doi.org/10.1063/1.1690231] [PMID: 15267776]

[24]   D. Almeida, F. Ferreira da Silva, J. Kopyra, G. García, and P. Limão-Vieira, "Anion formation in gas-phase potassium-uridine collisions", *Int. J. Mass Spectrom.,* vol. 365-366, pp. 243-247, 2014.
[http://dx.doi.org/10.1016/j.ijms.2014.01.023]

[25]   F. Ferreira da Silva, M. Lança, D. Almeida, G. García, and P. Limão-Vieira, "Anionic fragmentation of glycine upon potassium molecule collisions", *Eur. Phys. J. D,* vol. 66, pp. 78-84, 2012.
[http://dx.doi.org/10.1140/epjd/e2012-20751-y]

[26]   F. Ferreira da Silva, J. Rafael, T. Cunha, D. Almeida, and P. Limão-Vieira, "Electron transfer to aliphatic amino acids in neutral potassium collisions", *Int. J. Mass Spectrom.,* vol. 365-366, pp. 238-242, 2014.
[http://dx.doi.org/10.1016/j.ijms.2014.01.003]

[27]   P. D. Burrow, G. A. Gallup, A. M. Scheer, S. Denifl, S. Ptasinska, T. D. Märk, and P. Scheier, "Vibrational Feshbach resonances in uracil and thymine", *J. Chem. Phys.,* vol. 124, no. 124310, pp. 1-7, 2006.

[28]   A.W. Kleyn, J. Los, and E.A. Gislason, "Vibronic coupling at intersections of covalent and ionic states", *Phys. Rep.,* vol. 90, pp. 1-71, 1982.
[http://dx.doi.org/10.1016/0370-1573(82)90092-8]

[29]   A.W. Kleyn, and A.M.C. Moutinho, "Negative ion formation in alkali-atom-molecule collisions", *J. Phys. B,* vol. 34, pp. R1-R44, 2001.
[http://dx.doi.org/10.1088/0953-4075/34/14/201]

[30]   P. Limão-Vieira, A. M. C. Moutinho, and J. Los, "Dissociative ion-pair formation in collisions of fast potassium atoms with benzene and fluorobenzene", *J. Chem. Phys.,* vol. 124, no. 054306, pp. 1-10, 2006.

[31]   R. Antunes, *"PhD Thesis. The role of halouracils in radiotherapy studied by electron transfer in atom-molecule collisions experiments, Universidade Nova de Lisboa",*

[32]   R. Antunes, D. Almeida, G. Martins, N.J. Mason, G. García, M.J.P. Maneira, Y. Nunes, and P. Limão-

Vieira, "Negative ion formation in potassium-nitromethane collisions", *Phys. Chem. Chem. Phys.,* vol. 12, no. 39, pp. 12513-12519, 2010.
[http://dx.doi.org/10.1039/c004467a] [PMID: 20721400]

[33]   P. Limão-Vieira, F. Ferreira da Silva, and G.G. Goméz-Tejedor, *Electron transfer-induced fragmentation in (bio)molecules by atom-molecule collisions: negative ion formation; Radiation Damage in Biomolecular Systems.,* G.G. Gómez-Tejedor, M.C. Fuss, Eds., Springer: New York, 2012, pp. 59-70.
[http://dx.doi.org/10.1007/978-94-007-2564-5_3]

[34]   P. Limão-Vieira, "Private Communication",

[35]   D. Almeida, M-C. Bacchus-Montabonel, F. Ferreira da Silva, G. García, and P. Limão-Vieira, "Potassium-uracil/thymine ring cleavage enhancement as studied in electron transfer experiments and theoretical calculations", *J. Phys. Chem. A,* vol. 118, no. 33, pp. 6547-6552, 2014.
[http://dx.doi.org/10.1021/jp503164a] [PMID: 24818533]

[36]   S. Denifl, S. Ptasińska, G. Hanel, B. Gstir, M. Probst, P. Scheier, and T.D. Märk, "Electron attachment to gas-phase uracil", *J. Chem. Phys.,* vol. 120, no. 14, pp. 6557-6565, 2004.
[http://dx.doi.org/10.1063/1.1649724] [PMID: 15267547]

[37]   S. Denifl, F. Zappa, A. Mauracher, F. Ferreira da Silva, A. Bacher, O. Echt, T.D. Märk, D.K. Bohme, and P. Scheier, "Dissociative electron attachment to DNA bases near absolute zero temperature: freezing dissociation intermediates", *ChemPhysChem,* vol. 9, no. 10, pp. 1387-1389, 2008.
[http://dx.doi.org/10.1002/cphc.200800245] [PMID: 18551498]

[38]   F.A. Gianturco, F. Sebastianelli, R.R. Lucchese, I. Baccarelli, and N. Sanna, "Ring-breaking electron attachment to uracil: following bond dissociations via evolving resonances", *J. Chem. Phys.,* vol. 128, no. 17, p. 174302, 2008.
[http://dx.doi.org/10.1063/1.2913169] [PMID: 18465917]

[39]   T. Sommerfeld, "Electron-induced chemistry of 5-chlorouracil", *ChemPhysChem,* vol. 2, no. 11, pp. 677-679, 2001.
[http://dx.doi.org/10.1002/1439-7641(20011119)2:11<677::AID-CPHC677>3.0.CO;2-C]   [PMID: 23686903]

[40]   S. Denifl, S. Ptasinska, M. Probst, J. Hrusak, P. Scheier, and T.D. Märk, "Electron Attachment to the Gas-Phase DNA Bases Cytosine and Thymine", *J. Phys. Chem. A,* vol. 108, pp. 6562-6569, 2004.
[http://dx.doi.org/10.1021/jp049394x]

[41]   H. Abdoul-Carime, S. Gohlke, and E. Illenberger, "Site-specific dissociation of DNA bases by slow electrons at early stages of irradiation", *Phys. Rev. Lett.,* vol. 92, no. 16, p. 168103, 2004.
[http://dx.doi.org/10.1103/PhysRevLett.92.168103] [PMID: 15169265]

[42]   S. Ptasińska, S. Denifl, V. Grill, T.D. Märk, E. Illenberger, and P. Scheier, "Bond- and site-selective loss of H- from pyrimidine bases", *Phys. Rev. Lett.,* vol. 95, no. 9, p. 093201, 2005.
[http://dx.doi.org/10.1103/PhysRevLett.95.093201] [PMID: 16197213]

[43]   C. Desfrancois, H. Abdoul-Carime, S. Carles, V. Périquet, J.P. Schermann, D.M.A. Smith, and L. Adamowicz, "Experimental and theoretical ab initio study of the influence of N-methylation on the dipole-bound electron affinities of thymine and uracil", *J. Chem. Phys.,* vol. 110, pp. 11876-11883, 1999.
[http://dx.doi.org/10.1063/1.479175]

[44]   J.H. Hendricks, S.A. Lyapustina, H.L. de Clercq, J.T. Snodgrass, and K.H.J. Bowen, "Dipole bound, nucleic acid base anions studied via negative ion photoelectron spectroscopy", *Chem. Phys.,* vol. 104, pp. 7788-7791, 1996.

[45]   D. Almeida, D. Kinzel, F. Ferreira da Silva, B. Puschnigg, D. Gschliesser, P. Scheier, S. Denifl, G. García, L. González, and P. Limão-Vieira, "N-site de-methylation in pyrimidine bases as studied by low energy electrons and ab initio calculations", *Phys. Chem. Chem. Phys.,* vol. 15, no. 27, pp. 11431-11440, 2013.

[http://dx.doi.org/10.1039/c3cp50548k] [PMID: 23743926]

[46] I. Bald, J. Kopyra, and E. Illenberger, "Selective excision of C5 from D-ribose in the gas phase by low-energy electrons (0-1 eV): implications for the mechanism of DNA damage", *Angew. Chem. Int. Ed. Engl.,* vol. 45, no. 29, pp. 4851-4855, 2006.
[http://dx.doi.org/10.1002/anie.200600303] [PMID: 16819742]

[47] D. Almeida, R. Antunes, G. Martins, G. García, R.W. McCullough, S. Eden, and P. Limão-Vieira, "Mass spectrometry of anions and cations produced in 1–4 keV H−, O−, and OH− collisions with nitromethane, water, ethanol, and methanol", *Int. J. Mass Spectrom.,* vol. 311, pp. 7-16, 2012.
[http://dx.doi.org/10.1016/j.ijms.2011.11.009]

[48] P. Sulzer, S. Ptasinska, F. Zappa, B. Mielewska, A.R. Milosavljevic, P. Scheier, T.D. Märk, I. Bald, S. Gohlke, M.A. Huels, and E. Illenberger, "Dissociative electron attachment to furan, tetrahydrofuran, and fructose", *J. Chem. Phys.,* vol. 125, no. 4, p. 44304, 2006.
[http://dx.doi.org/10.1063/1.2222370] [PMID: 16942139]

[49] C. Winstead, and V. McKoy, "Low-energy electron scattering by deoxyribose and related molecules", *J. Chem. Phys.,* vol. 125, no. 7, p. 074302, 2006.
[http://dx.doi.org/10.1063/1.2263824] [PMID: 16942334]

[50] B.C. Ibănescu, O. May, and M. Allan, "Cleavage of the ether bond by electron impact: differences between linear ethers and tetrahydrofuran", *Phys. Chem. Chem. Phys.,* vol. 10, no. 11, pp. 1507-1511, 2008.
[http://dx.doi.org/10.1039/b718130b] [PMID: 18327306]

CHAPTER 4

On the Sudden Removal of Two Outer-shell Electrons in Atoms

A.C. F. Santos[1,*] and **D. P. Almeida**[2]

[1] *Instituto de Física, Universidade Federal do Rio de Janeiro, PO 68528, 21941-972 Rio de Janeiro, RJ, Brazil*

[2] *Departamento de Física, Universidade Federal de Santa Catarina, Florianópolis, 88040-900 – Brazil*

Abstract: Electron correlation (EC) in atoms is at the same time very important and challenging investigation topic due to its many-body interactions, nevertheless it may be the main process to describe the absolute electron impact cross section for the ionization of atoms and molecules. The outer-shell double photoionization of a multi-electron target is usually a less important process in comparison to ionization of a single electron by a photon and it is determined completely by electron-electron interaction. The objective in this chapter has been upon a review of the recently found relationship between SO amplitudes and the number of target electrons as well as their density (electrons per volume). Through the comparison between the asymptotic values of the experimental ratios of the double-to-single photoionization cross-sections from the literature, a scaling law was achieved. This scaling allows us to predict the SO amplitudes for several atomic elements up to xenon within a factor two. The electron SO amplitudes following outer-shell photoionization have been plotted as a function of the target atomic number, Z, and static polarizability, α^x . Our results are in qualitative agreement with the experimental data.

I. INTRODUCTION

Throughout outer shell double ionization (DPI) by a single photon, electrons from the atom have a small but not zero probability to be ejected to be ionized. The process can be represented as

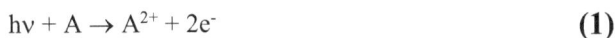

$$h\nu + A \rightarrow A^{2+} + 2e^- \tag{1}$$

where hv is the incident photon energy, A is the atomic target and 2e⁻ represents the two ejected electrons.

* **Corresponding author A.C.F. Santos:** Instituto de Física, Universidade Federal do Rio de Janeiro, PO 68528, 21941-972 Rio de Janeiro, RJ, BrazilS; Tel: +55 21 3938-7947; E-mail: toni@if.ufrj.br

Two distinct mechanisms commit to it. These mechanisms, which differ in nature, are often denoted as the two-step one interaction (TS-1 or KO, knock-out) and shake-off (SO) [1 - 10], (see Fig. **1**). Both mechanisms are given much consideration in trying to comprehend double ionization in atoms and molecules. Electron correlations (*i.e.*, electron-electron interaction) are many-body effects and can be defined as the difference between the experimental and the prediction from the independent particle approximation. It should be observed that multiple ionization can take occur also under multi-photon photoionization, as for instance, by using intense short wavelength beams from a free electron laser.

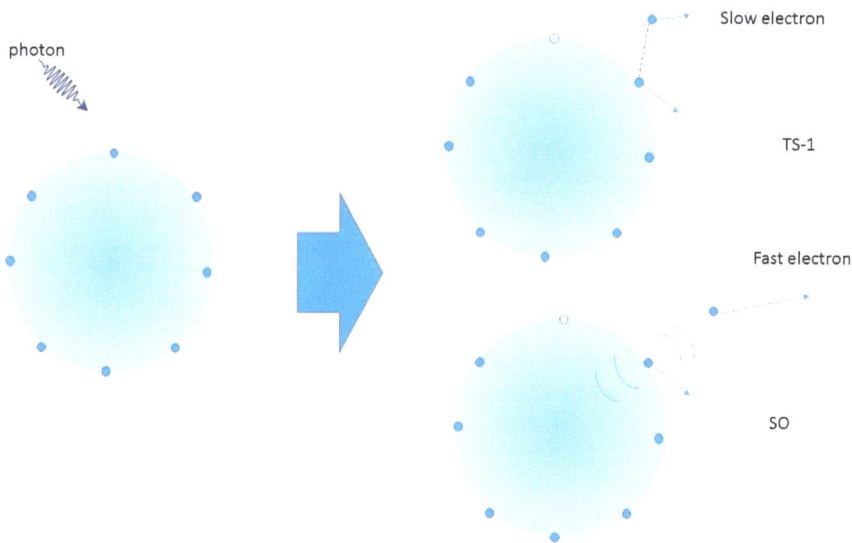

Fig. (1). Illustration of the shake-off (SO) and two-step 1 (TS-1) mechanisms in double photoionization of the valence shell. In the TS-1 (or knock-out) process, the outgoing (slow) photoelectron knocks out the second electron. The TS-1 process is of classical nature, while the SO process is of quantum nature.

The ratio of double-to-single ionization cross section of atoms by a single photon approaches a constant value which can be explained in the framework of a sudden approximation by the SO mechanism. Shake-off is well described in the high-energy limit, but its theoretical definition at low and intermediate photon energies is not clear and has been the subject of discussion. Actually, several shake-off definitions can be found in the literature [1 - 7]. Generally, different shake-off definitions approach the correct asymptotic value at high photon energies. Nevertheless, at low and intermediate photon energies, they may diverge significantly. Therefore, to date, there is no unique definition for shake-off at low and intermediate photon energies. Notwithstanding, in order to connect the theoretical results with the experimental data, some considerations have been paid to the problem of which mechanisms govern double ionization and in what energy ranges the contributing mechanisms are significant [1 - 10].

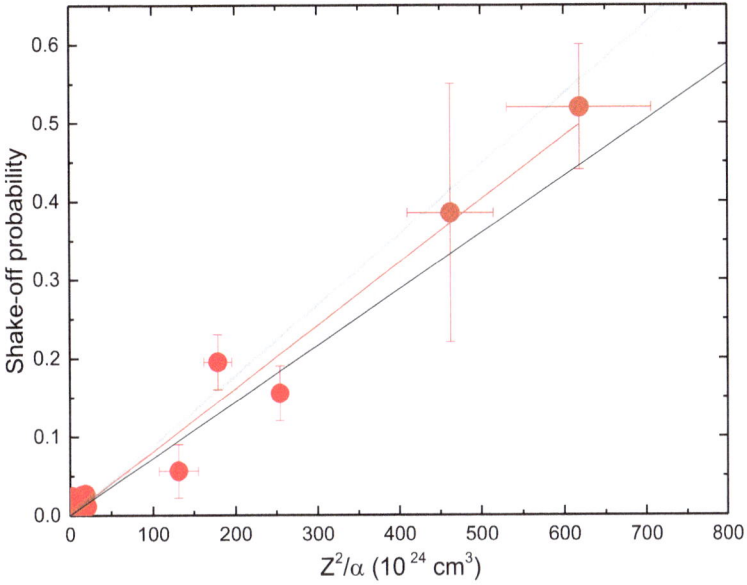

Fig. (2). Experimental shake off probabilities as a function of $Z^2/\alpha\infty$. The shaded area shows the upper and lower limits for the SO probability.

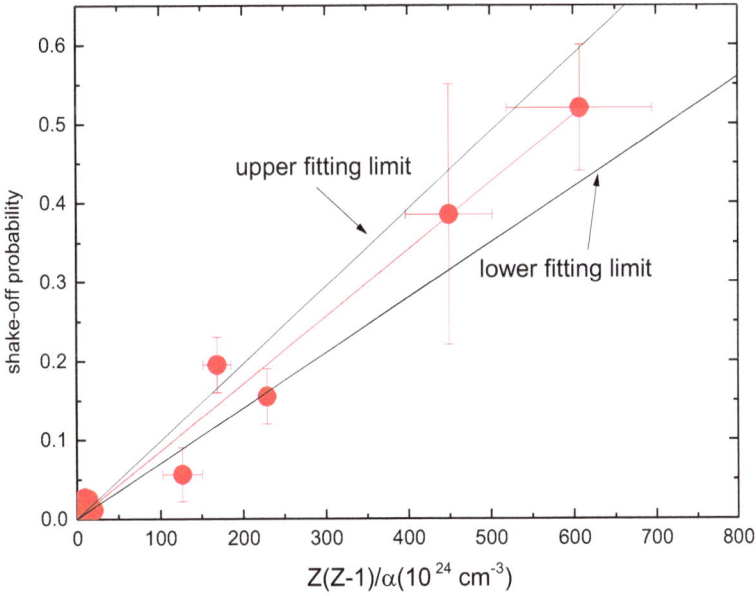

Fig. (3). Experimental shake off probabilities as a function of $Z(Z-1)/\alpha\infty$.

Being aware of electron correlation is a leading task of paramount importance to atomic and molecular physics, chemistry, solid state physics, and many other areas, see for instance ref [11]. On the theoretical standpoint, a truthful description of EC in many-electron systems still represents a difficult task. In the case of the shake-off mechanism, in which case the photon energy is far above the double ionization potential, the photoelectron quickly flies way from the target, without any collision with the rest of the target electrons. Then, the remaining target electrons are exposed to a sudden change in the Coulomb interaction and can be shaken-up (excited) or shaken-off (ejected to continuum). In the sudden approximation (SA) method, the shake-off (SO) contribution is calculated by the overlap between the initial and final target states. As follows, the SO probability is independent of the ionizing agent and its energy.

Several papers aiming the double photoionization have focused on the helium atom, which is the smaller system where double ionization can take place. Double-photoionization (DPI) data for high-Z atoms other than the rare gas atoms are still uncommon due to the complexity of understanding the data ([12, 13], for instance). In addition, theoretical studies are also mainly devoted to the helium atom. The energy dependence of the shake-off mechanism has been well determined. The shake-off probability is zero for photon energies lower than the double ionization potential. It then increases up to a constant value at energies well above the double ionization threshold, *i.e.*, the photon energy above which the shake-off probability is constant. The perturbative nature of the shake-off permit us to use the sudden approximation for photons at high energies. The sudden approximation (SA) is the simplest way to compute shake-off transitions. Shake-off is usually portrayed as a two-step process. In the first step, the photoelectron is ejected, while in the second step the rest of the target electrons feel a drastic change in the Coulombic potential so that the projection of the initial atomic wave function on the assembly of final wave functions is responsible for the shake-off amplitudes. The total shake-off probability, is given by [11]

$$P_{SO} = 1 - \prod_{i}^{core} \left[\int_{0}^{\infty} \psi^{f}{}_{n_i l_i}(r) \psi_{n_i l_i}(r) dr \right]^2 \qquad (2)$$

Where, the second term symbolizes the probability that each of the residual electrons with initial wave function $\psi_{n_i l_i}(r)$ "overlaps" into final wave functions $\psi^{f}{}_{n_i l_i}(r)$ with the same quantum numbers.

In very interesting paper, T. D. Thomas [2] assumed that the time evolution of the

Coulomb potential shift as a result of a vacancy formation may be ascribed by the Gauss error function and solved the time-dependent Schrodinger equation. By adopting some simple hypothesis, the shake-off transition probability can be written as

$$P_{SO}(\varepsilon) = P_{SO}(\infty)e^{-\left(\frac{m_e r^2 I_B^2}{2\hbar^2 \varepsilon}\right)} \tag{3}$$

Where $P_{SO}(\infty)$ is the shake-off probability at the sudden limit $(t \to \infty)$, r is the atomic radius, and I_B is the binding energy of the shaken electron. For an incident photon with energy E, the excitation energy is given by $\varepsilon = E - I_B$.

To evaluate the effect of outer shell electrons and the relative role of initial-state and final-state electron correlation for the double photoionization (DPI), an empirical shake-off/knock-out model established on a hypothesis of incoherent sum of the double-to-single cross section ratios for the shake-off and knock-out processes was suggested [10]. Within that picture, the ratio of double-to-single ionization as a function of the photon energy can be written as:

$$P_{DPI} = \frac{\sigma^{2+}}{\sigma^+} = P_{KO} + P_{SO} \tag{4}$$

Pattard [14] suggested a universal shape function for single and multiple photoionization that postulates a very good parameterization of photoionization cross sections for double ionization of He-like ions. The analytical formula is given by:

$$\sigma^{2+}(\varepsilon) = \sigma_{max}^{2+} \left(\frac{\varepsilon}{\varepsilon_{max}}\right)^{\alpha} \left[\frac{\alpha + 7/2}{\alpha \frac{\varepsilon}{\varepsilon_{max}} + 7/2}\right]^{\alpha + 7/2} \tag{5}$$

Where, ε_{max} is the excess energy where the cross section is maximum, σ_{max}^{2+} is the value of the maximum cross section, and $\alpha = 1.056$ (do not confuse with the polarizability).

Attempts to understand double ionization results have tried in the literature by a considerable number of authors. A scaling model for the ratio of double-to-single ionization of low-Z atoms has been suggested by Bluett *et al.* [15]. Bluett *et al.* tried to build a model that could forecast the double-to-single photoionization ratios for high-Z elements. Nonetheless, their model has been applied only to a few elements to date.

Not long ago, DuBois, Santos, and Manson [9], examined published experimental results for cross-section ratios for double ionization of low-Z rare gases (He, Ne, and Ar). The cross sections corresponding to collisions with antiprotons and bare ions ranging from Z=1 to Z = 92, were understood according to the first- (SO) and second-order (two-step 2) interference model suggested by McGuire [16]. The authors published a full analysis of existing experimental electron removal cross sections and cross section ratios for an extensive range of systems and impact energies. Their analysis supports the McGuire model [16] for double electron removal. By using experimental data on single and double electron ionization to obtain second-order contributions to double ionization, analytical formulae for the first and second order mechanisms giving rise to single and double ionization were obtained. Multiply charged ion data were used to obtain projectile charge dependences while p^+, p^- and multicharged ion cross sections for helium, neon, and argon atoms were used to obtain scaling dependences. The data suggest that the set of these scaling gives rise to a reasonable compliance for double outer-shell electron ionization of any atomic system by any bare projectile for most impact energies. The formulae and scaling presented by the authors that were extracted from experimental data can be used to pathfind theoretical calculations of double electron removal.

Lately, Alcantara *et al.* [10] studied the significance of the shake-off and knock-out mechanisms for the double photoionization of the dichloromethane (CH_2Cl_2) molecule. The probabilities for SO and KO mechanisms associated with outer-shell photoionization were obtained as a function of the photon wavelength.

With the intention of examine more deeply the dependence of the SO probability on the electron density and target atomic number, in this chapter we present studies on the saturation effect of the shake process in double ionization. The shake-off amplitudes after valence double ionization for atoms, from He to Xe, were obtained from published data. The scope of this chapter is to some shed light on the main origin of the dependence of the shake-off probability on the target parameters.

II. METHODS

The scaling laws for SO probabilities for atoms were inferred in the following method. Initially, by using experimental data found in the literature, a set of experimental shake-off probabilities (see Table **1**) was built for atomic numbers in the range from He, Z=2 to Xe, Z=54. Then, the dependences of the shake-off probabilities on target atomic number, Z, and target polarizability,α, were methodically investigated. At last, the achieved functions were merged and adjusted in pursuance of to give rise to a best fit for all targets.

Table 1. Experimental data.

Element	Atomic number	Shake-off probability	References for SO probability	Polarizability (Å^3)	References for polarizability
He	2	0.015 0.0376	[17, 34]	0.2050 0.22	[22, 23]
Li	3	0.01081*	[22]	24.3 15.96	[22, 23]
Be	4	0.0214* 0.026	[33]	5.60 7.77	[22, 23]
Ne	10	0.16 0.09 0.15	[17, 27, 28]	0.3956 0.39	[22, 23]
Na	11	0.01117* 0.0128	[17, 34]	24.08 18.67	[22, 23]
Mg	12	0.0083* 0.011	[17, 31]	10.6 14.13	[22, 23]
Ar	18	0.21 0.18 0.19	[17, 27, 28]	1.6411 1.98	[22, 23]
K	19	0.0112*	[34]	43.4 37.58	[22, 23]
Ca	20	0.035* 0.015	[31, 32]	22.8 25.0 33.80	[22, 23]
Zn	30	0.09* 0.022	[17, 33]	5.75 5.6 8.12	[22, 23]
Kr	36	0.39 ± 0.20	[17, 25, 27, 32]	2.4844 3.12	[22, 23]
Xe	54	0.5 ± 0.1	[17, 25, 26]	4.044 5.38	[22, 23]

* Data obtained by rough estimative from low energy data, where SO does not dominate.

In looking for the scaling, some basic assumptions were admitted. Firstly, following Alcantara *et al.* [10], the shake off contributions should be an increasing function of the electron density. Then, several authors point out that the shake-off should be higher for high-Z targets. Next, the shake-off contribution falls if the target electrons are more deeply bound, *i.e.*, with the amount of energy necessary to ionize any specific electron. In addition, a considerable number of papers have been concerned with the interdependences between atomic polarizability, α, and atomic or molecular volume and/or ionization potential. Lastly, the bottom line criteria is that the shake-off interdependences ought to be a simple function and be valid for a large collection of atoms.

In order to establish the set of experimental shake-off amplitudes, the principal sources for double photoionization of valence electrons were the complete work from reference [17] and the review of Wehlitz [18]. The shake-off probabilities were methodically arranged in terms of target atomic number, Z, and polarizability, α. The data were plotted to check for scaling parameters dependences. The present study encloses eleven atomic targets from helium to xenon. The data for this work are listed in Table **1**. The experimental shake-off values in the literature are different by a factor up to two, while the polarizabilities are in agreement within 25%.

Initially, one focuses our attention to the electron correlation (EC) with respect to the atomic number Z. Electron correlation is a measure of spatial correlation, since this interaction is a function of the distance between the electrons through the Coulomb field. The electronic coulombic interactions generate a medium potential, and so a specific electron is affected by the remaining electrons of the system. Consequently, the more electrons the atom has, the larger is the change in the electron state. For a neutral atom, the number of electrons is equal to its atomic number, Z. Hence, it would be expected that the dot product of the initial electron state on the multitude of final states (Eq. 1) should also be a function of Z. Moreover, a target electron correlates with the other *(Z-1)* electrons. If we add up these interactions over all *Z* electrons, it is expected that electron correlation would depend on Z as *Z(Z-1) Z^2*.

Then again, in the case of the CH_2Cl_2 molecule (fourty-eight electrons), its lower shake-off probability, in connection to argon atoms (eighteen electrons), was accredited to the fewer many-electron correlations due to its bulkier volume [10]. Thus, the larger volume, the weaker is the electron-electron correlations because of the 1/r Coulomb potential involved. The atomic volume, is commonly seen as a critical factor of electrical polarizability [19] that is a kind of electron correlation, however, in this case, from electrons associated with a different particle. The dynamical polarizability $α_λ$ is given by [20 - 30]:

$$\alpha_\lambda = E_H^2 a_o^3 \int_{I_p}^{\infty} \frac{\left(\dfrac{df}{dE}\right)}{\left(E^2 - \varepsilon^2\right)} dE \qquad (6)$$

Where ε is the photon energy, $E_H = 27.21$ eV, is the Hartree energy, and df/dE is photoabsorption differential oscillator strength (PDOS). The PDOS and the photoabsortion cross section are related quantities as given by

$$\sigma(Mb) = 109.75 \frac{df}{dE} (eV^{-1}) \qquad (7)$$

The experimental values of the dipole polarizability are generally derived [ref] from dielectric constant or refractive index type measurements. The static dipole polarizability is given by S(-2) sum rule [20]:

$$S(-2)a_o^3 = \alpha_\infty = \lim_{\lambda \to \infty} \alpha_\lambda \qquad (8)$$

where

$$S(-2) = \int_{I_p}^{\infty} \left(\frac{E}{E_H}\right)^{-2} \left(\frac{df}{dE}\right) dE \qquad (9)$$

Therefore, it should not be unexpected that the shake-off probability should also depend on target volume. In the case of a uniform conducting sphere of volume V, the polarizability α^∞, is given strictly in terms of the volume [19, 21]:

$$\alpha_\infty = \frac{3V}{4\pi} \qquad (10)$$

Firsthand dependence between atomic and molecular polarizabilities and volumes have been described by several papers [19, 21]. The polarizability α^∞ provides knowledge of how intense the electronic cloud can be modified by an external electric field. From the arguments presented above, the shake-off contribution should scale to the electron density (Z/V) times a function of Z, $f(Z)$. Thus, the shake-off probability should be proportional to

$$P_{SO} \sim \begin{cases} \dfrac{Z(Z-1)}{\alpha_{\infty}} \\ or \\ \dfrac{Z^2}{\alpha_{\infty}} \end{cases} \tag{11}$$

III. RESULTS AND DISCUSSIONS

By means of the data of Table **1** and the assumed connection between the shake-off probability measured values for the shake-off amplitudes for the double photoionization of atomic systems ranging from helium (Z=2) to xenon (Z=54). The uncertainties were calculated by bearing in mind the experimental errors bars as well as the uncertainties associated to the atomic polarizabilities from references [22] and [23]. A straight line dependence is observed. The equations 12 and 13 are fitted, weighted with respect to the size of the error bars that provides a method of estimating SO amplitudes for valence double ionization for atomic systems.

$$P_{SO}(\infty) = b\left(\frac{Z^2}{\alpha_{\infty}}\right) \tag{12}$$

$$P_{SO}(\infty) = c\left(\frac{Z^2 - Z}{\alpha_{\infty}}\right) \tag{13}$$

where b=$(8.2 \pm 0.4)\times 10^{-24}$ cm^3 and c=$(8.5 \pm 0.4)\times 10^{-24}$ cm^3.

Of specific attention are the rare gases which are closed shell atoms. Figs. (**4** and **5**) shows the shake-off probability as a function of Z^2/α for noble gases only. In this case, the fitting gives b = $(8.1 \pm 0.9)\times 10^{-24}$ cm^{-3} and c=$(8.5\pm1.0)\times 10^{-24}$ cm^3. The bunch of data points at the left end of Figs. (**2** and **3**) are the atoms in group 2 of the periodic table (Ca, Mg, Be, *etc.*), which show lower shake-off probabilities. The data were obtained by means of extrapolation from low photon energy data, and should be considered as a naive estimative. The shake-off contribution for the atoms in group two of the periodic table and the He atom are virtually the same within the associated uncertainties.

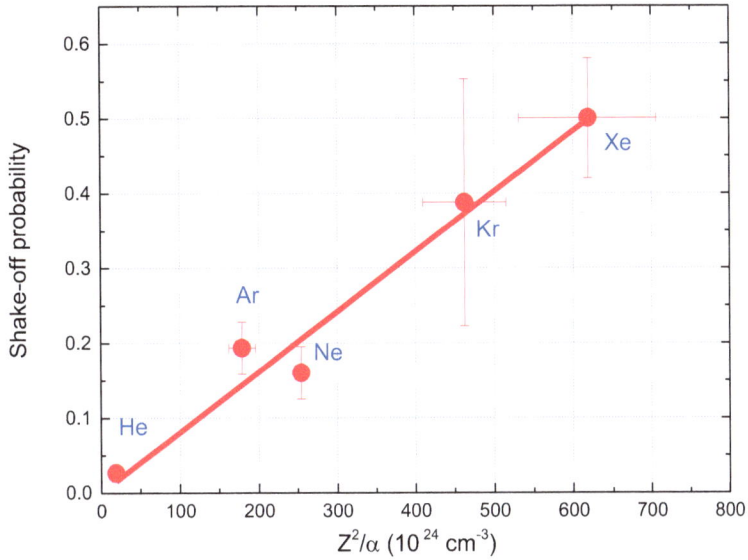

Fig. (4). The same of Fig. 2 for noble gases only.

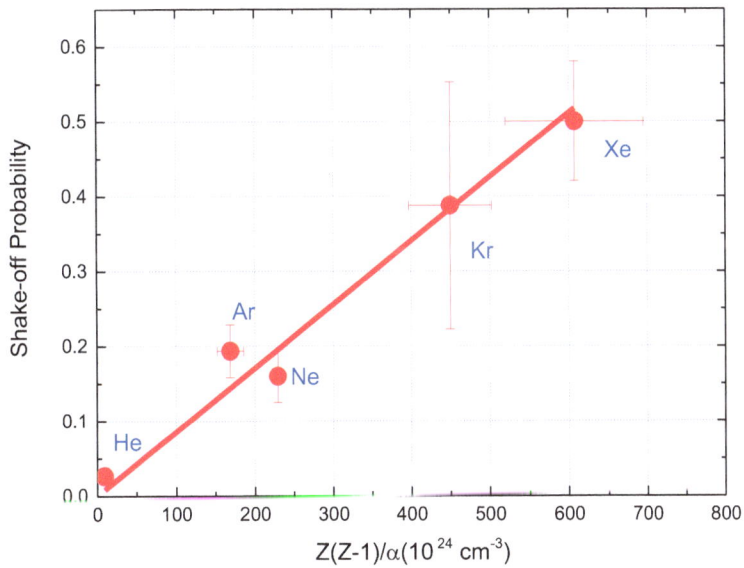

Fig. (5). The same of Fig. 3 for noble gases only.

IV. SUMMARY, CONCLUDING REMARKS, AND OUTLOOK

In conclusion, understanding electron correlations is not one of the most important issues in atomic and molecular physics, but it is also fundamental for an accurate theoretical description of many-electron atomic systems were determined using experimental data measured for a broad variety of atoms from the literature. It was found that the shake-off amplitudes scale roughly as Z^2/α^x, where Z is the target atomic number and α^x is the target static electric polarizability. The present scaling is estimated to have forecast ability with accuracy within factor of two. Single-photon double ionization processes can also produce core-shell hollow atoms and molecules (*i.e.* systems with an empty innermost shell and occupied outer shells) and is a growing field which allows to investigate experimentally ultrafast electron dynamics within atoms and molecules.

CONSENT FOR PUBLICATION

Not applicable.

CONFLICT OF INTEREST

The author (editor) declares no conflict of interest, financial or otherwise.

ACKNOWLEDGEMENT

This work was supported in part by CNPq (Brazil).

REFERENCES

[1]   T Åberg, "Correlation Energy of K-shell Electrons", *Phys. Rev/,* vol. 162, p. 5, 1967.
        [http://dx.doi.org/10.1103/PhysRev.162.5]

[2]   T.D. Thomas, "Transition from Adiabatic to Sudden Excitation of Core Electrons", *Phys. Rev. Lett.,* vol. 52, p. 417, 1984.
        [http://dx.doi.org/10.1103/PhysRevLett.52.417]

[3]   Ki. Hino, T. Ishihara, F. Shimizu, N. Toshima, and J.H. McGuire, "Double photoionization of helium using many-body perturbation theory", *Phys. Rev. A,* vol. 48, no. 2, pp. 1271-1276, 1993.
        [http://dx.doi.org/10.1103/PhysRevA.48.1271] [PMID: 9909732]

[4]   T. Pattard, and J. Burgd"orfer, "Half-collision model for multiple ionization by photon impact", *Phys. Rev. A,* vol. 64, p. 042720, 2001.
        [http://dx.doi.org/10.1103/PhysRevA.64.042720]

[5]   A. Kheifets, "On different mechanisms of the two-electron atomic photoionization", *J. Phys. At. Mol. Opt. Phys.,* vol. 34, p. L247, 2001.
        [http://dx.doi.org/10.1088/0953-4075/34/8/102]

[6]   T.Y. Shi, and C.D. Lin, "Double photoionization and transfer ionization of he: shakeoff theory revisited", *Phys. Rev. Lett.,* vol. 89, no. 16, p. 163202, 2002.
        [http://dx.doi.org/10.1103/PhysRevLett.89.163202] [PMID: 12398719]

[7]   T. Schneider, P.L. Chocian, and J.M. Rost, "Separation and identification of dominant mechanisms in double photoionization", *Phys. Rev. Lett.,* vol. 89, no. 7, p. 073002, 2002.
        [http://dx.doi.org/10.1103/PhysRevLett.89.073002] [PMID: 12190518]

[8]   A.C.F. Santos, A. Hasan, T. Yates, and R.D. DuBois, "Doubly differential measurements for multiple ionization of argon by electron impact: Comparison with positron impact and photoionization", *Phys. Rev. A,* vol. 67, p. 052708, 2003.
        [http://dx.doi.org/10.1103/PhysRevA.67.052708]

[9]   R.D. DuBois, A.C.F. Santos, and S.T. Manson, "Empirical formulas for direct double ionization by bare ions: Z = −1 to 92", *Phys. Rev. A,* vol. 90, p. 052721, 2014.
[http://dx.doi.org/10.1103/PhysRevA.90.052721]

[10]   K.F. Alcantara, A.H.A. Gomes, W. Wolff, L. Sigaud, and A.C.F. Santos, "Outer-shell double photoionization of CH2Cl2", *Chem. Phys.,* vol. 429, p. 1, 2014.
[http://dx.doi.org/10.1016/j.chemphys.2013.11.019]

[11]   F. Martin, P.D. Burrow, Z. Cai, P. Cloutier, D. Hunting, and L. Sanche, "DNA Strand Breaks Induced by 0–4 eV Electrons: The Role of Shape Resonances", *Phys. Rev. Lett.,* vol. 93, p. 068101, 2004.

[12]   D.P. Almeida, M.A. Scopel, R.R. Silva, and A.C. Fontes, "Single and double ionisation of mercury by electron impact", *Chem. Phys. Lett.,* vol. 341, pp. 5-6, 490, 2001.
[http://dx.doi.org/10.1016/S0009-2614(01)00520-6]

[13]   D.P. Almeida, and R.R. Silva, "Multiple ionization of mercury near threshold by electron impact", *J. Elect. Spec. Rel. Phen.,* vol. 107, no. 3, p. 205, 2000.
[http://dx.doi.org/10.1016/S0368-2048(00)00120-1]

[14]   T. Pattard, "A shape function for single-photon multiple ionization cross sections", *J. Phys. At. Mol. Opt. Phys.,* vol. 35, p. L207, 1984.
[http://dx.doi.org/10.1088/0953-4075/35/10/103]

[15]   J.B. Bluett, D. Lukic, S.B. Whitfield, and R. Wehlitz, "Double photoionization near threshold", *Nucl. Instrum. Methods Phys. Res. B,* vol. 241, p. 114, 2005.
[http://dx.doi.org/10.1016/j.nimb.2005.07.015]

[16]   J.H. McGuire, "Double ionization of helium by protons and electrons at high velocities", *Phys. Rev. Lett.,* vol. 49, p. 1153, 1982.
[http://dx.doi.org/10.1103/PhysRevLett.49.1153]

[17]   N. Saito, and I.H. Suzuki, "Shake-off processes in photoionization and Auger transition for rare gases irradiated by soft X-rays", *Phys. Scr.,* vol. 49, p. 80, 1994.
[http://dx.doi.org/10.1088/0031-8949/49/1/011]

[18]   R. Wehlitz, Simultaneous emission of multiple electrons from atoms and molecules using synchrotron radiation.*Atomic, Molecular, and Optical Physics* vol. 58. , 2010, pp. 1-76.
[http://dx.doi.org/10.1016/S1049-250X(10)05806-4]

[19]   P. Politzer, P. Jin, and J.S. Murray, "Atomic polarizability, volume and ionization energy", *J. Chem. Phys.,* vol. 117, p. 8197, 2002.
[http://dx.doi.org/10.1063/1.1511180]

[20]   R.A. Bonham, "On the validity of using the TKR sum rule to place zero angle electron impact spectra on an absolute scale", *Chem. Phys. Lett.,* vol. 31, p. 559, 1975.
[http://dx.doi.org/10.1016/0009-2614(75)85085-8]

[21]   T. Brinck, J.S. Murray, and P. Politzer, "Polarizability and volume", *J. Chem. Phys.,* vol. 98, p. 4305, 1993.
[http://dx.doi.org/10.1063/1.465038]

[22]   D.R. Lide, *CRC Handbook of Chemistry and Physics* 89th ed. CRC Press: New York, 2008-2009.

[23]   S. Karwowski, J. Saxena, K.M.S Fraga., *Handbook of Atomic Data* Elsevier Scientific Publishing Company, 1976.

[24]   T.A. Carlson, C.W. Nestor, N. Wasswerman, and J.D. McDowell, "Calculated ionization potentials for multiply charged ions", *At. Data,* vol. 2, p. 63, 1970.
[http://dx.doi.org/10.1016/S0092-640X(70)80005-5]

[25]   Th.M. El-Sherbini, and M.J. van der Wiel, "Oscillator strengths for multiple ionization in the outer and first inner shells of Kr and Xe", *Physica,* vol. 62, p. 119, 1972.
[http://dx.doi.org/10.1016/0031-8914(72)90154-1]

[26]   M.Y. Adam, Ph. D. Thesis, Université de Paris-Sud, (1978).

[27]   D.M.P. Holland, K. Codling, J.B. West, and G.V. Marr, "Multiple photoionisation in the rare gases from threshold to 280 eV", *J. Phys. B,* vol. 12, p. 2485, 1979.
[http://dx.doi.org/10.1088/0022-3700/12/15/008]

[28]   G. R. Wight, "Oscillator strengths for double ionization in the outer shells of He, Ne and Ar", *J. Phys. B,* vol. 9, p. 1319, 1976.
[http://dx.doi.org/10.1088/0022-3700/9/8/016]

[29]   E. Murakami, T. Hayaishi, A. Yagishita, and Y. Morioka, "Multiple and partial photoionization cross sections in the Kr 3d ionization region", *Phys. Scr.,* vol. 41, p. 468, 1990.
[http://dx.doi.org/10.1088/0031-8949/41/4/019]

[30]   R. Wehlitz, and S.B. Whitfield, "Valence double photoionization of beryllium", *J. Phys. At. Mol. Opt. Phys.,* vol. 34, pp. L719-L725, 2001.
[http://dx.doi.org/10.1088/0953-4075/34/22/103]

[31]   A. S. Kheifets, "Valence-shell double photoionization of alkaline-earth-metal atoms", *Phys. Rev. A,* vol. 75, p. 042703, 2007.
[http://dx.doi.org/10.1103/PhysRevA.75.042703]

[32]   A.S. Kheifets and I. Bray., "Valence-shell double photoionization of alkaline-earth-metal atoms", *Phys. Rev. A,* vol. 75, p. 042703, 2007.

[33]   D.M.P. Holland, K. Codling, and J.B. West, "Near threshold double photoionisation in Zn, Cd and Hg", *J. Phys. B,* vol. 8, no. 15, p. 1473, 1982.

[34]   P.N. Juranić, J. Nordberg, and R. Wehlitz, "Single- and double-photoionization data of Na and K corroborate the existence of a universal scaling law for the ratio", *Phys. Rev. A,* vol. 74, p. 042707, 2006.
[http://dx.doi.org/10.1103/PhysRevA.74.042707]

CHAPTER 5

Multielectronic Processes in Particle and Antiparticle Collisions with Rare Gases

Claudia C. Montanari[*]

Instituto de Astronomía y Física del Espacio, Consejo Nacional de Investigaciones Científicas y Técnicas, and Universidad de Buenos Aires, casilla de correo 67, sucursal 28, C1428EGA, Buenos Aires, Argentina

Abstract: In this chapter we analyze the multiple ionization by impact of $|Z|=1$ projectiles: electrons, positrons, protons and antiprotons. Differences and similarities among the cross sections by those four projectiles allows us to have an insight on the physics involved. Mass and charge effects, energy thresholds, and relative importance of collisional and post-collisional processes are discussed. For this purpose, we performed a detailed theoretical-experimental comparison for single up to quintuple ionization of Ne, Ar, Kr and Xe by particles and antiparticles. We include an extensive compilation of the available data for the sixteen collisional systems, and the theoretical cross sections by means of the continuum distorted wave eikonal initial state approximation. We underline here that post-collisional ionization is decisive to describe multiple ionization by light projectiles, covering almost the whole energy range, from threshold to high energies. The normalization of positron and antiproton measurements to electron impact ones, the lack of data in certain cases, and the future prospects are presented and discussed.

Keywords: Antiparticle, Antiproton, Electron, Ionization, Multiple ionization, Noble gases, Positron, Proton, Particle, Ne, Ar, Kr, Xe.

INTRODUCTION

Multiple ionization is a challenging subject, which plays an important role in the knowledge of many-electron processes, such as multiple-electron transitions, collisional and post-collisional ionization or electron correlation effects. The goal of this contribution is to deepen in the study of the multielectronic processes by collision of $|Z|=1$ particles and antiparticles. For this purpose, we focused in the multiple ionization of the heaviest rare gases, Ne, Ar, Kr and Xe. We analyzed the differences and similarities in the cross sections of equal-charge *versus* equal

[*] **Corresponding author Claudia C. Montanari:** Instituto de Astronomía y Física del Espacio, Consejo Nacional de Investigaciones Científicas y Técnicas, and Universidad de Buenos Aires, casilla de correo 67, sucursal 28, C1428EGA, Buenos Aires, Argentina; Tel: +54 11 5285 7860; E-mail: mclaudia@iafe.uba.ar

Antônio Carlos Fontes dos Santos (Ed.)

mass projectiles. The ionization cross sections by light (electrons and positrons) and heavy (protons and antiprotons) projectiles are quite different in the low and intermediate energy regions. On the contrary, in the high energy region all these cross sections converge. However, this convergence to proton impact values is different for antiprotons, positrons or electrons. A detailed knowledge of this tendency from intermediate to high impact energies is important because of the experimental normalization of relative cross sections of antiparticles to electron or to proton values. Classical reviews on particle and antiparticle collisions are also available [1 - 5].

The theoretical description of the multiple ionization processes by these projectiles must consider the charge and mass effects, the projectile trajectories, and the energy thresholds. The post-collisional ionization (PCI) due to Auger-type processes following inner-shell ionization, enhance the final number of emitted electrons. For heavy projectiles this is important at high energies. For light projectiles, such as electrons and positrons, PCI dominates the highly-charged ion production in the whole energy range, even close to the energy threshold. The ionization by proton and electron impact has been studied since the early years of the development of atomic physics. However, the experimental data on multiple ionization by high energy protons could not be theoretically described until the last fifteen years [8 - 12]; and by electron impact only recently [13]. This was possible by a consistent inclusion of the branching ration for PCI within the independent electron model.

Multiple ionization is also a sensitive test for the experimental work. Measurements require highly advanced techniques to get all possible channels and final states. For protons, they must separate pure ionization from capture channels, which enhance the data in the intermediate energy region [6]. In the case of positron impact, the total ionization values at low energies include positronium formation [7]. On the other hand, the higher the order of ionization the smaller the cross sections. The quintuple ionization cross sections of Kr and Xe are of the order of the 10^{-19} cm^2 at high energies, while for Ar they are 10^{-20} cm^2 or even 10^{-21} cm^2.

The published experimental data of multiple ionization of Ne, Ar, Kr and Xe by proton impact is profuse [6, 8, 9, 14 - 17, 19, 20]. Much more abundant are the measurements by electron impact. They include the pioneering works by Schram and co-workers in the 60s [21] to the present [22 - 32].

Instead, the experimental data by antiparticle is more scarce, what is reasonable. Antiprotons are produced in high-energy physics sources and then decelerated for atomic collisions experiments. This research has been developed by Knudsen and coworkers at CERN [16, 33 - 37], first in the low energy antiproton ring (LEAR) and nowadays in the antiproton decelerator (AD). A very recent state *of art* of antiproton impact ionization has been published by Kirchner and Knudsen [38]. In the case of positrons, most of the experimental publications report only values of single ionization [39 - 43], some articles include double or triple ionization measurements [44 - 51], and only one paper quadruple ionization [49]. It is worth to mention that antiproton and positron values are normalized to electron impact ionization cross sections at high energies.

In this chapter we present a comparison of the multiple ionization cross sections by impact of electrons, positrons, protons and antiprotons. We consider the rare gases Ne, Ar, Kr and Xe, and final charge states from +1 to +3 (Ne), and +5 (Ar, Kr and Xe). The comparison includes the theoretical results obtained by employing the continuum distorted wave eikonal initial state approximation in [12, 13, 52, 53] and the available experimental measurements. The extensive compilation of data for the four projectiles and the four targets included here, allow us to have a wide vision of the experimental *state of art*, and some future prospects.

THEORETICAL DESCRIPTION

The theoretical description for multiple ionization reviewed in this chapter relies on three approximations:

1. **The independent particle model** (IPM): the ejected electrons ignore each other, neglecting the correlation in the final state and the changes in the target potential due to the successive loss of electrons. Under this assumption, the probability of multiple ionization can be expressed as a multinomial combination of independent ionization probabilities.

2. **The continuum distorted wave eikonal initial state** (CDW-EIS) approximation [54], [55]. This is a proved model to describe intermediate and high energy multiple ionization by protons and antiprotons [12, 52]. In [13, 53], the CDW-EIS is adapted to describe ionization by electrons or positrons, by taking into account the finite momentum transferred, the non-linear trajectory and the mass effect. Light particle ionization is characterized by the sharp energy thresholds, which are different for single to quintuple ionization. The results in [53] include the different thresholds of energy within the multinomial expansion following [56].

3. **The post-collisional ionization- (PCI), independent of the projectile.** Thus, it is included within the multinomial expression in a semiempirical way following [8, 10], using the experimental branching ratios of the charge-state distribution after a single initial vacancy. Present formalism is explained in

[12]. A detailed compilation of these branching ratios is available in [13].

In what follows the implications of these approximations on the calculations are summarized. More details of them can be found in the mentioned references.

THE INDEPENDENT PARTICLE MODEL FOR MULTIPLE IONIZATION

Within the IPM, the probability of direct ionization of exactly q_j electrons of the j subshell as a function of the impact parameter b, $P_{(q_j)}(b)$, is obtained as a multinomial distribution of the ionization probabilities $p_j(b)$ given by

$$P_{(q_j)}(b) = \binom{N_j}{q_j} \left[p_j(b)\right]^{q_j} \left[1 - p_j(b)\right]^{N_j - q_j} \tag{1}$$

where N_j is the total number of electrons in the subshell. If n electrons are ionized from the different shells, $n = \sum_j q_j$, the total probability of direct ionization is

$$P_{(n)}(b) = \sum_{q_1 + q_2 + \ldots = n} \prod_j P_{(q_j)}(b) \tag{2}$$

and the cross section corresponding to the direct ionization of exactly n electrons is

$$\sigma_N = \int P_{(n)}(b) 2\pi b \, db \tag{3}$$

The main and more sensitive parameter are the ionization probabilities as function of the impact parameter, $p_j(b)$. Different models have been applied to multiple ionization calculations in the last fifteen years by Montenegro and coworker [8, 9, 12, 57 - 60], Kirchner *et al..* [10, 61 - 65], Rivarola *et al.* [11, 66, 67], and Miraglia and collaborators [12, 13, 52, 53, 68, 69].

CDW-EIS Ionization Probabilities by Proton, Antiproton, Electron and Positron Impact

We review here the results for proton, antiproton, electron and positron impact multiple ionization in [12, 13, 52, 53]. These results employ the CDW-EIS code by Miraglia [54]. Details of these calculations are in [52, 54]. The aim is to discuss the scope and limitations of this model to deal with particle-antiparticle and heavy-light projectile effects.

The CDW-EIS ionization probabilities $p_j(b)$ as function of the impact parameters are the seeds to be introduced in the multinomial expression given by (1). As expected, these results tend to the first Born approximation ones at sufficiently high impact energies [52]. In fact, the extension of the theoretical results for impact velocities $v > 8$ a.u. in [12, 13, 52, 53] is achieved using the first Born approximation.

On the other hand, the light particle calculations were performed adapting the CDW-EIS for equal-charged light projectiles, *i.e.* protons → positrons, antiprotons → electrons. This calculation accounted for the finite momentum transferred, the non-linear trajectory (very different for positrons and electrons due to the repulsive or attractive potentials), and the minimum energy. A detailed explanation and the corresponding equations can be found in [13].

The CDW-EIS results in [12, 13, 52, 53] describe *pure* ionization of the target. No electron transfer or transfer followed by projectile electron loss is included. These processes are possible only for positive projectiles, and may enlarge the experimental cross sections at low energies.

Ionization processes present a sharp energy threshold: ionization is not possible for projectile energies below the binding energy of the target outer electrons. The threshold for each shell of electrons must be taken into account within the multinomial expression (1). One of the differences between light and heavy projectiles is that, on equal impact velocity, light projectiles have impact energies of the order of the target binding energies or even lower. This is the main cause of differences between ionization cross sections by light and heavy projectiles at intermediate to low energies.

Experimentally the threshold or appearance energy is well known and measured in electron and positron-impact ionization. For n-fold ionization the experiments indicate that this appearance energy is much greater than n times the binding energy of the outermost electrons. This was analyzed in [56] considering not only the energy gap, but also the mean velocity of the outgoing electrons in semi-classical way. This proposal was included in the multiple ionization calculations by electron and positron impact in [53]. However, it must be mention that even though the threshold itself is rather well described; the CDW-EIS approximation fails at low impact energies, *i.e.* when the energy loss by the projectile is comparative to the impact energy. We will return to this later in this chapter.

Auger Type Postcollisional Ionization

The PCI is due to the rearrangement of the target atom after the ionization, ending in a highly charged ion. We consider here the different processes (shake off, Auger decay and emission) involving the emission of one or more electrons long after the collision. Following Cavalcanti *et al.* [8] and Spranger and Kirchner [10], the PCI is included inside the binomial equation (1) in a semi-empirical way by using experimental branching ratios of charge state distribution after single photoionization. The deeper the initial hole, the greater the number emitted electron in PCI.

The branching ratios $F_{j,k}$ verify the unitary condition, $\sum_{k=0}^{k_{max}} F_{j,k} = 1$, with k being the number of electrons emitted in PCI after the single ionization of the j subshell [12]. This fact is employed to introduce them in (1) as follows

$$P_{(q_j)}(b) = \binom{N_j}{q_j} \left[p_j(b) \sum_{k=0}^{k_{max}} F_{j,k} \right]^{q_j} \left[1 - p_j(b) \right]^{N_j - q_j} \tag{4}$$

Afterwards, the addition of probabilities is rearranged in order to put together those terms that contribute to the same number of final emitted electrons. This rearrangement gives rise to new probabilities of exactly n emitted electrons, including direct ionization and PCI, $P_{(n)}^{PCI}$ (see section 2.3 in [12] for details). Thus, the corresponding cross sections of n-fold including PCI are

$$\sigma_n^{PCI} = \int P_{(n)}^{PCI}(b) 2\pi b \, db \tag{5}$$

RESULTS AND DATA OF PARTICLE AND ANTIPARTICLE IONIZATION: CHARGE AND MASS EFFECTS

What follows is a revision of the theoretical CDW-EIS results for multiple ionization of Ne, Ar, Kr and Xe by protons and antiprotons [52], electrons [13], and positrons [53]. This is achieved by comparing and analyzing the cross sections for the four $|Z|=1$ projectiles together, and contrasting them with the experimental data available in the literature. The comparison is performed on equal impact velocities, and is plotted as a function of the corresponding electron impact energy. The change to proton impact energies is straightforward, just an m_p/m_e factor (the ratio of masses of heavy and light projectiles).

In the following sections the results for Ne, Ar, Kr and Xe are presented. We will note that the threshold for light particles is quite well described by the theory. However, we will focus on the description of the intermediate to high impact energy processes that is the range of validity of the CDW-EIS.

THE MULTIPLE IONIZATION OF NE

In Figs. (**1-3**) we plotted together the CDW-EIS results for proton, antiproton, electron, and positron impact ionization [13, 52, 53], for final Ne^+ to Ne^{3+}. As mentioned in [60] the multiple ionization of Ne with ten bound electrons cannot be extended within the IPM to higher final charge states without taking into account the changes in the Ne potential. On the contrary, for targets with larger number of bound electrons, the IPM is expected to work.

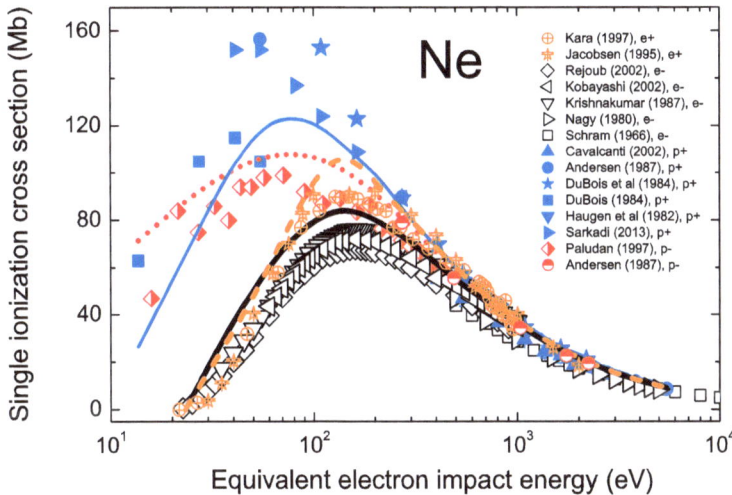

Fig. (1). Single ionization of Ne by |Z|=1 projectiles as function of the impact energy, considering equal velocity heavy and light particles. Curves: CDW-EIS results for proton (blue thin solid-line), antiproton (red dotted-line), electron (black thick solid-line) and positron (orange dashed-line) impact. Symbols: details in the inset; the references are: for protons p+ [6, 8, 14 - 16, 20], for electrons e- [21, 23 - 26, 29, 30] for positrons e+ [34, 39], and for antiprotons p- [16, 48].

Ne Single Ionization

In Fig. (**1**) we include the theoretical curves and the experimental measurements for the single ionization cross sections of Ne by protons (p+), antiprotons (p-), electrons (e-), and positrons (e+). It can be observed that the main difference between heavy and light projectiles are the lower values at intermediate and low energies, and the sharp threshold around 23 eV. Instead, all the results converge at high energies. The intermediate energy region is the most interesting one. We note that for equal mass, the values for projectiles with charge Z=-1 are below the Z=+1 ones. It could be said that the ionization of Ne is more effective by positive charges.

The theoretical description is rather good for the heavy projectiles. In the case of protons, below the maximum the agreement is fine only with the data by DuBois [6] who separates pure ionization and capture. This may be the difference between DuBois [6] data and the recent Sarkadi *et al.* [20] experimental values at intermediate to low energies. In the case of antiprotons we remark the good agreement with Andersen measurements at high energies [16]. Instead, the data by Paludan [35] are somewhat below the predictions, but the tendency is correct. For single-ionization by light particles, the theoretical values overestimate around the maximum, and describe nicely the measurements at high energies.

The normalization of antiparticle cross sections is a point of discussion [53]. The experimental cross sections by positrons and antiprotons are mostly relative values normalized to well known and tested high energy electron impact data, such as Rapp *et al.* [70], Sorokin *et al.* [71], or Krishnakumar *et al.* [24]. For example, the positron data by Kara *et al.* [45] and antiproton data by Paludan *et al.* [35] are both normalized to electron impact data by Krishnakumar *et al.* [24] at 800-1000 eV. The comparison displayed in Fig. (**1**) shows that for the theoretical model, antiprotons and positrons tend faster to proton values than to electron ones at high energies. At 400 eV proton, antiproton and positron curves are together and 10% above electron impact values. Even around 1000 eV the electron single ionization is still a little below the rest. This implies that antiproton and positron relative values could be normalized to total proton-impact ionization cross sections, such as those by Rudd *et al.* [72], at no so high energies.

Ne Double Ionization

In Fig. (**2**) we display the double ionization of Ne. We use the logarithmic scale to emphasize the fall down and convergence at high energies, much more drastic than for single-ionization.

The double ionization cross sections by antiprotons at intermediate to low velocities, *i.e.* v < 2.3 a.u. is higher than by proton impact. The inverse than for single ionization. And this reversal can be seen in the data and in the theoretical description too. Another point to remark in Fig. (**2**) is that theoretically, above 200 eV (electron impact) the proton, antiproton and even positron curves converge to a single value. Instead, for electron impact the convergence is above 600 eV. Experimentally, the positron data are below the electron measurements. It is not clear how the normalization of the single ionization positron measurements to electron data may weigh at this level.

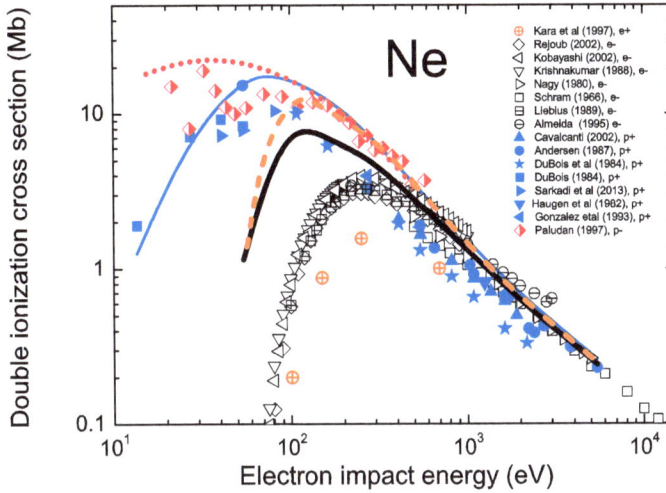

Fig. (2). Double ionization of Ne by |Z|=1projectiles as function of the impact energy, considering equal velocity heavy and light particles. Curves: CDW-EIS results for proton (blue thin solid-line), antiproton (red dotted-line), electron (black thick solid-line) and positron (orange dashed-line) impact. Symbols: details in the inset; the references are: for protons p+ [6, 8, 14 - 16, 20], for electrons e- [21, 23 - 26, 29, 30] for positrons e+ [39], and for antiprotons p- [48].

At high energies the description is actually good. Our values include the post-collisional Auger and shake off processes. At 3 keV the PCI contribution is 60$\%$ of the total double ionization (see Fig. **2** in [53]). At intermediate energies this model clearly overestimates the electron and positron-impact data. This is related to the inclusion of PCI due to shake-off of the outer shell electrons of Ne, as explained in [13]. This is an interesting open topic: The inclusion of PCI allows us to describe the high energy region, but in the case of Ne, it is produces too high values at intermediate energies. On the contrary, if only direct double ionization is considered, the maximum for electron impact cross section and the energy threshold are better described.

Ne Triple Ionization

The triple ionization cross sections of Ne are displayed in Fig. (**3**). The theoretical results for proton and antiproton triple ionization are rather good. They clearly show that antiprotons produce larger triple ionization for energies around the maximum than protons. Note that this is just the inverse for single ionization.

The comparison of the cross sections by heavy and light projectiles shows the expected lower values for light projectiles and the convergence at high energies. The CDW-EIS results for light projectiles overestimate the measurements near the energy threshold. This is partially due to the overestimation of the shake-off contribution, and partially due to being in the limit of validity of the model, as mentioned previously in this chapter.

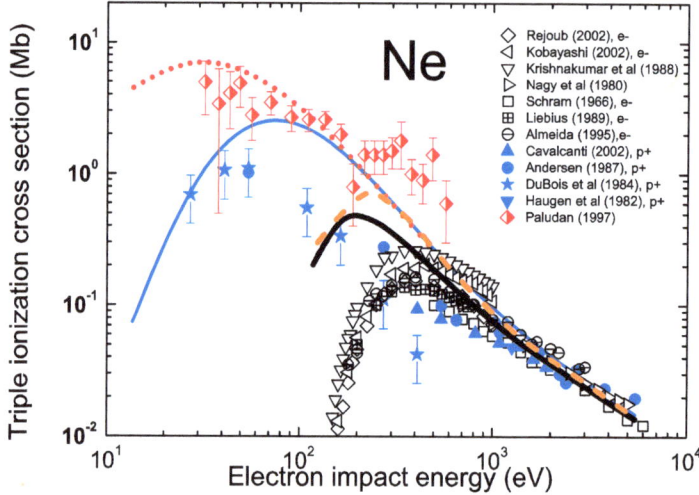

Fig. (3). Triple ionization of Ne by |Z|=1 projectiles as function of the impact energy, considering equal velocity heavy and light particles. Curves: CDW-EIS results for proton (blue thin solid-line), antiproton (red dotted-line), electron (black thick solid-line) and positron (orange dashed-line) impact. Symbols: details in the inset; the references are: for protons p+ [6, 8, 14 - 16, 20], for electrons e- [21, 23 - 26, 29, 30], and for antiprotons p- [48].

It is very interesting to mention that the high energy data could not be described if the post-collisional contribution to triple ionization (initial ionization of inner shells and rearrangements processes) is not included [13]. This contribution is important above 400 eV, being 90% of the triple ionization at 1 keV.

THE MULTIPLE IONIZATION OF AR

We analyze the Ar multiple ionization, from single to quintuple. In Figs. (**4-8**) we displayed together the CDW-EIS results for proton, antiproton, electron and positron in [52, 53], [13],positron, and an updated compilation of experimental data for the four projectiles.

In the following figures it will be noted that the theoretical description shows better agreement with the measurements than for Ne. This behavior is enhanced for heavier targets, as we will observe for Kr and Xe in the following sections. The IPM works better to describe the multiple ionization in multielectronic targets, so that the number of loose electrons is much smaller than the number of bound electrons.

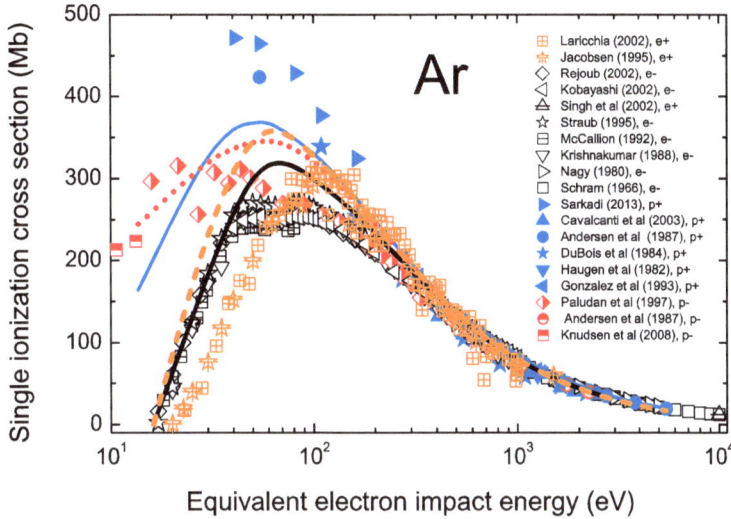

Fig. (4). Single ionization of Ar by |Z|=1 projectiles as function of the impact energy, considering equal velocity heavy and light particles. Curves: CDW-EIS results for proton (blue thin solid-line), antiproton (red dotted-line), electron (black thick solid-line) and positron (orange dashed-line) impact. Symbols: details in the inset; the references are: for protons p+ [6, 8, 14 - 16, 20], for electrons e- [21, 23 - 26, 29, 30] for positrons e+ [34, 36], and for antiprotons p- [16, 48, 49].

Ar Single Ionization

In Fig. (**4**) the single ionization of Ar by the four |Z|=1 projectiles is displayed. The good description of the experimental values by the CDW-EIS can be noted. The proton impact measurements by Andersen *et al.* [16] and by Sarkadi *et al.* [20] at impact velocities below 2.5 a.u. are clearly higher than the theoretical values. It may be possible that these single ionization measurements by proton impact include capture, enhancing the number of measured Ar^+. Obviously, the antiproton measurements are pure ionization, and capture is not possible.

In the high energy region, the convergence of the proton, antiproton, positron and electron measurements and also of the theoretical curves, is clearly above 300 eV. Again, our model predicts that the ionization by positrons tends to antiproton and proton values at lower energies than to the electron impact values. In the case of Ar, this tendency is also found in the experimental data by Laricchia and coworkers [42, 40].

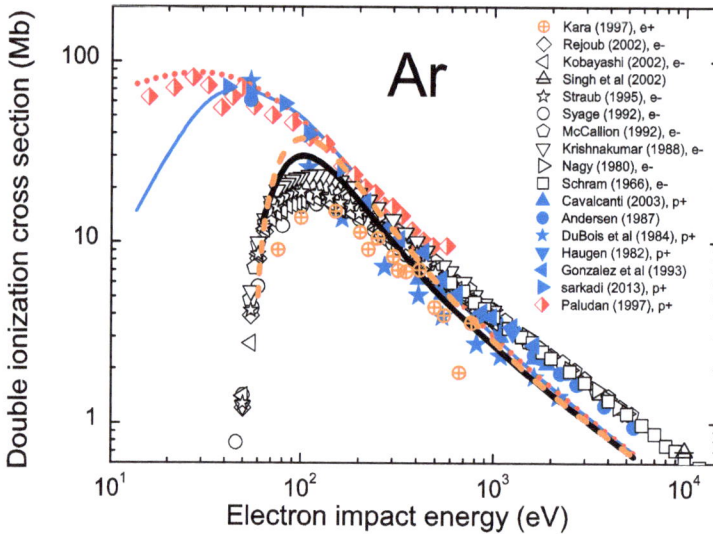

Fig. (5). Double ionization of Ar by |Z|=1 projectiles as function of the impact energy, considering equal velocity heavy and light particles. Curves: CDW-EIS results for proton (blue thin solid-line), antiproton (red dotted-line), electron (black thick solid-line) and positron (orange dashed-line) impact. Symbols: details in the inset; the references are: for protons p+ [5], Cavalcanti02 [9, 14 - 18], gonzalez, for electrons e- [21, 23 - 26, 29, 30] for positrons e+ [39], and for antiprotons p- [48].

Ar Double Ionization

In Fig. (**5**) the double ionization of Ar is shown. The general description obtained with the CDW-EIS for the heavy and the light projectiles is good. Though this model is not capable of describing the low energy processes, the results are quite good in this region too. The threshold for double ionization of Ar by electron and positron impact is correctly described, as compared to the data.

At high energies $E > 1\ keV$ our curves are below most of the electron-impact and proton-impact data. It is worth to mention that all the Ar shells have been included in this calculation, even the K-shell. So this unexplained undervalue of the theory may be related to the empirical branching ratios employed.

Ar Triple Ionization

In Fig. (**6**) the triple ionization of Ar is displayed. The experimental data for heavy projectiles is rather well described, though overestimated around the maximum. The theory predicts antiproton values above the proton ones at the maximum of the cross section. Experimentally, the data for protons match well with the data for antiprotons.

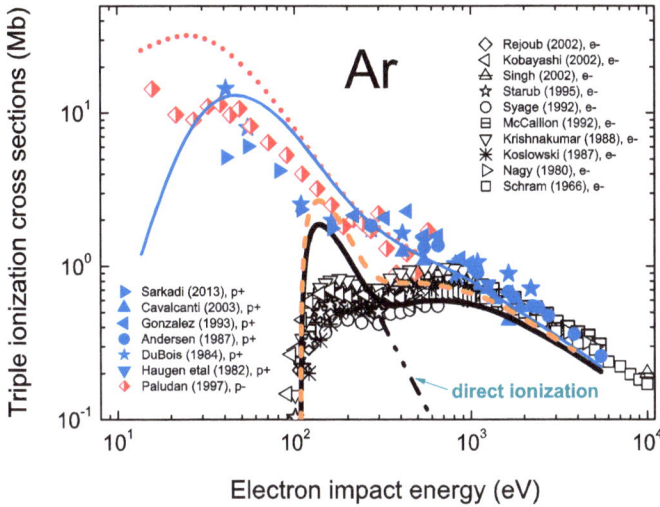

Fig. (6). Triple ionization of Ar by |Z|=1 projectiles as function of the impact energy, considering equal velocity heavy and light particles. Curves: CDW-EIS results for proton (blue thin solid-line), antiproton (red dotted-line), electron (black thick solid-line) and positron (orange dashed-line) impact; black dashed-double-dotted line, direct triple ionization by electron impact. Symbols: details in the inset; the references are: for protons p+ [9, 14 - 16], gonzalez, sarkadi, for electrons e- [21, 23 - 26, 29, 30] and for antiprotons p- [48].

In this figure we remark the importance of the PCI by including the direct triple ionization values (dashed double-dotted line) and the triple ionization including PCI (solid line) for electron impact ionization. As can be noted, PCI is correctly included. It is interesting to note a double-shoulder shape in the triple ionization by light projectiles. This is due to the passage from the energy region of direct ionization (only valence-shell ionization) to the region of higher energies, where PCI is not negligible (inner shell ionization). Note that this double shoulder can be observed also in the electron impact measurements. The theory reproduces the shape but overestimates below 250 eV. It would be interesting to have measurements of triple ionization by positron impact to study this effect too.

Ar Quadruple Ionization

In Fig. (**7**) the quadruple ionization cross sections of Ar are displayed. Again we emphasize the importance of PCI by including the direct quadruple ionization for the electron-impact case (dashed double-dotted curve). Above 300 eV the PCI is crucial, instead the direct ionization falls down, being negligible for impact energies above 400 eV. The theory predicts a small shoulder around 200 eV related to these two different ionization mechanisms. It can also be noted that the threshold for electron-impact quadruple ionization is nicely described.

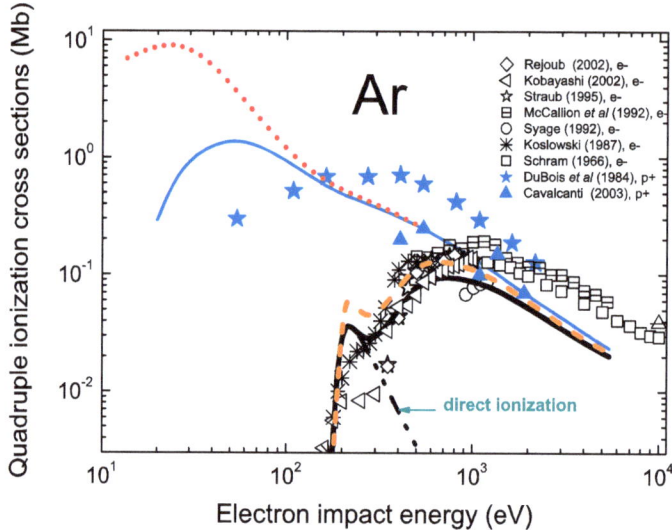

Fig. (7). Quadruple ionization of Ar by |Z|=1 projectiles as function of the impact energy, considering equal velocity heavy and light particles. Curves: CDW-EIS results for proton (blue thin solid-line), antiproton (red dotted-line), electron (black thick solid-line) and positron (orange dashed-line) impact. Symbols: details in the inset; the references are: for protons p+ [9, 15], and for electrons e- [21, 23 - 26, 29, 30].

Fig. (**7**) shows that there are no antiparticle data for quadruple ionization of Ar. This is related to the difficulties to measure such low values, *i.e.* around or less than 10^{-18} cm^2 for proton impact, around or less than 10^{-19} cm^2 for electron impact.

Ar Quintuple Ionization

Finally in Fig. (**8**) the quintuple ionization of Ar is displayed. The only data available in the literature is for electron-impact quintuple ionization. Nevertheless we include the CDW-EIS results for the four |Z|=*1* projectiles. Hopefully, these cross sections could be tested with future measurements by proton, antiproton and positron impact.

The theoretical description is very good considering it is quintuple ionization of Ar in an IPM approximation. The prediction of the threshold at 230 eV is also fine. In this figure, the cross sections by electron and positron impact are entirely due the PCI following inner-shell ionization (1s, 2s, and 2p). And this is valid in the whole energy range, even just above the threshold. The direct-ionization contribution is so small that it is out of scale in Fig. (**8**), *i.e.* less than 2×10^{-22} cm^2.

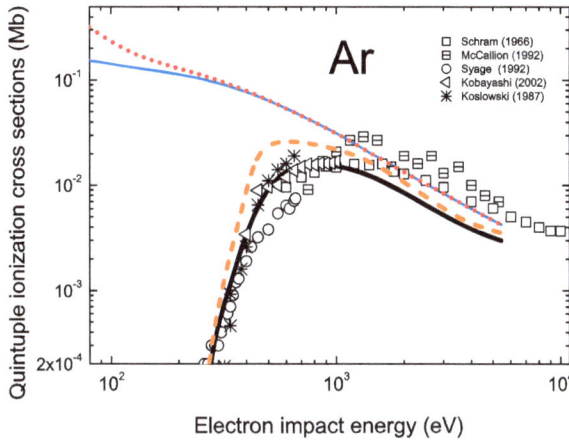

Fig. (8). Quintuple ionization of Ar by |Z|=1projectiles as function of the impact energy, considering equal velocity heavy and light particles. Curves: CDW-EIS results for proton (blue thin solid-line), antiproton (red dotted-line), electron (black thick solid-line) and positron (orange dashed-line) impact. Symbols: details in the inset; the references for electrons e- [21, 23 - 26, 29, 30].

THE MULTIPLE IONIZATION OF KR

We present in this section a comparative study of Kr multiple ionization by the four |Z|=1 projectiles, covering from single up to quintuple ionization. In Figs. (9-13) we displayed together the CDW-EIS results for proton, antiproton, electron and positron [52, 13, 53], and the available experimental data.

Kr Single Ionization

In Fig. (9) we display the single ionization cross sections of Kr. The theoretical-experimental agreement is very good. Note that for electron impact the theoretical description agrees in the maximum and in the energy threshold. Above 100 eV all the data seem to converge. Similarly to the case of Ar, the data by Sarkadi *et al.* [20] for proton impact are quite above the CDW-EIS predictions. But in this case we can also compare with the measurements by DuBois and coworkers [6, 17]. DuBois experiments separate capture and pure ionization. The differences between Sarkadi *et al.*. [20] and DuBois data [6, 17] below 60 eV are clear. This seems to confirm the presence of capture in [20].

Kr Double Ionization

In Fig. (10) we can observe and analyze the double ionization of Kr by particle and antiparticle impact. The good description of Kr double ionization by antiproton, proton and electron impact is remarkable. The theoretical and experimental values show clearly that the cross sections for heavy projectiles are quite similar for impact velocities above 1.7 a.u.

The theoretical results show that the double ionization cross sections by the light projectiles are much lower than those of the heavy projectiles at intermediate energies, with the threshold around 40 eV. These values reproduce the electron impact data rather well. However, the positron data by Kara *et al.* [45] are below the theoretical predictions and quite below the electron data.

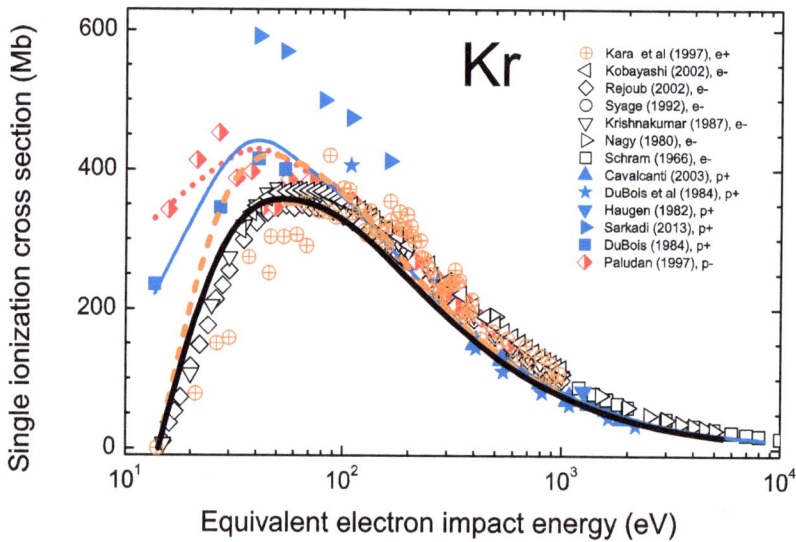

Fig. (9). Single ionization of Kr by |Z|=1projectiles as function of the impact energy, considering equal velocity heavy and light particles. Curves: CDW-EIS results for proton (blue thin solid-line), antiproton (red dotted-line), electron (black thick solid-line) and positron (orange dashed-line) impact. Symbols: details in the inset; the references are: for protons p+ [6, 8, 14 - 16, 20], for electrons e- [21, 23 - 26, 29, 30] for positrons e+ [39], and for antiprotons p- [48].

At low impact energies the mass-effect dominates. The cross sections by light projectiles are below the heavy projectiles ones, with the characteristic sharp threshold. This contrasts with the charge-effect at high energies. Above 100 eV positron values are closer to proton values than to electron ones. The high energy convergence of the different curves and data is evident at high energies, as expected.

Kr Triple Ionization

The theoretical results presented in Fig. (**11**) nicely describe the measurements of proton, antiproton and electron impact. No measurements for triple ionization of Kr by positrons have been reported yet. For light projectiles, PCI is the main contribution almost in the whole energy range. This is remarked in Fig. (**11**) with a separate curve that shows the direct triple ionization without PCI. Again, the small hump near the threshold is associated with the appearance of the PCI contribution. It is present in the theoretical curve for electron impact, and may be noted in the certain data. The bigger hump for positron impact could be an artifact of the theoretical calculations. However, no theoretical experimental comparison is possible for positron impact.

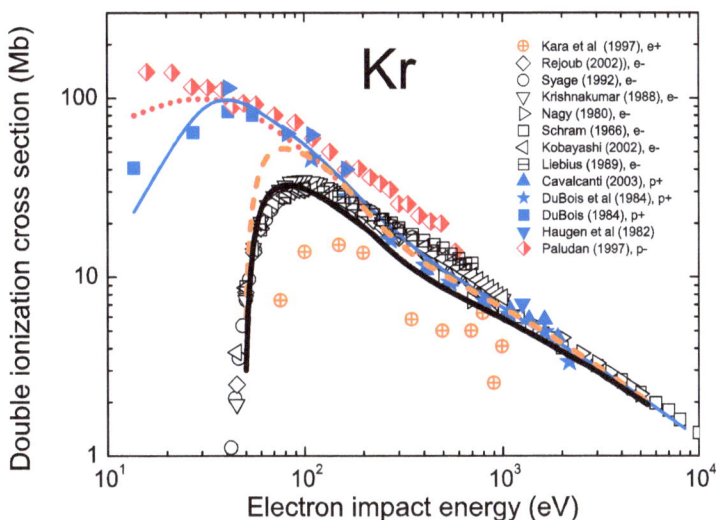

Fig. (10). Double ionization of Kr by |Z|=1 projectiles as function of the impact energy, considering equal velocity heavy and light particles. Curves: CDW-EIS results for proton (blue thin solid-line), antiproton (red dotted-line), electron (black thick solid-line) and positron (orange dashed-line) impact. Symbols: details in the inset; the references are: for protons p+ [6, 8, 14 - 16], [20], for electrons e- [21, 23 - 26, 29, 30] for positrons e+ [39], and for antiprotons p- [48].

As already mentioned, the mass effect is more important than the charge effect at low and intermediate energies, while equal charge determines the energy for which the curves converge. Theoretically, the positron cross sections tend to proton ones for $E > 1 \, keV$. Instead electron impact cross sections converge to the rest above 2 keV. This may indicate that when the ionization of inner shells dominates, the mass effect prevails over the charge effect.

Kr Quadruple Ionization

Fig. (**12**) shows the quadruple ionization of Kr. The theoretical-experimental agreement is noticeable. The comparison is very interesting. The direct ionization is a minor contribution for light particles even near the energy threshold. In the case of heavy particles PCI is important at rather low energies. This is remarked in Fig. (**12**) by including the theoretical direct quadruple ionization cross section by proton impact. Below 700 eV the theoretical results for positron impact are close to the electron ones, but no comparison with the experiments is possible for positron impact. The lack of antiparticle cross sections is evident in this figure. We expect that the present study of interest for future measurements.

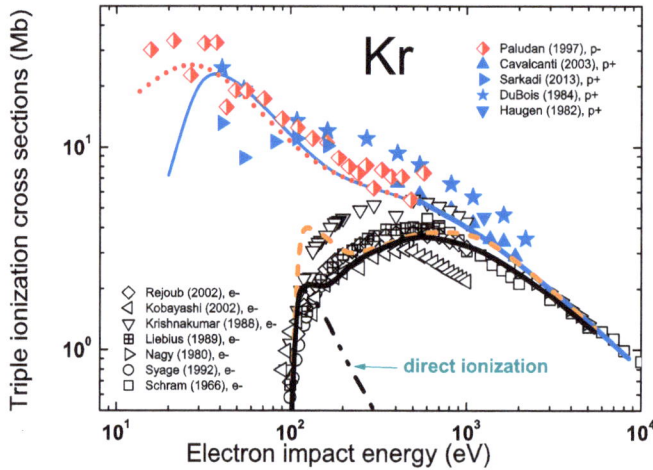

Fig. (11). Triple ionization of Kr by |Z|=1 projectiles as function of the impact energy, considering equal velocity heavy and light particles. Curves: CDW-EIS results for proton (blue thin solid-line), antiproton (red dotted-line), electron (black thick solid-line) and positron (orange dashed-line) impact. Symbols: details in the inset; the references are: for protons p+ [6, 8, 14 - 16, 20], for electrons e- [21, 23 - 26, 29, 30] and for antiprotons p- [48].

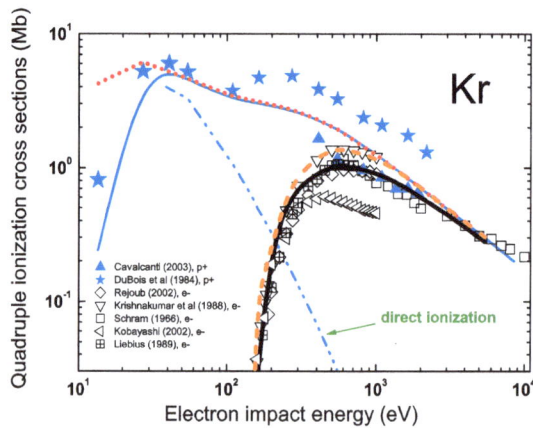

Fig. (12). Quadruple ionization of Kr by |Z|=1 projectiles as function of the impact energy, considering equal velocity heavy and light particles. Curves: CDW-EIS results for proton (blue thin solid-line), antiproton (red dotted-line), electron (black thick solid-line) and positron (orange dashed-line) impact. Symbols: details in the inset; the references are: for protons p+ [6, 8, 14 - 16, 20], and for electrons e- [21, 23 - 26, 29, 30].

Kr Quintuple Ionization

Finally, the quintuple ionization is displayed in Fig. (**13**). The behavior is similar to the one observed for quadruple ionization. Note that for proton impact there is only one set of experimental data, by DuBois [6]. The case is different for electron impact measurements, which are by far the most abundant in all the cases studied here. However, we included the theoretical predictions for the four |Z|=1projectiles in view of future measurements.

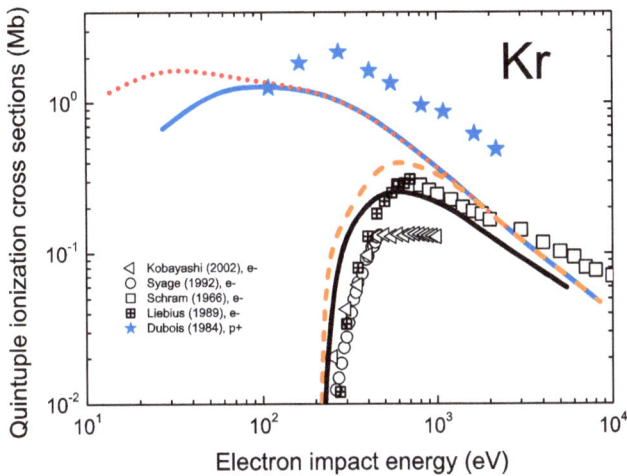

Fig. (13). Quintuple ionization of Kr by |Z|=1projectiles as function of the impact energy, considering equal velocity heavy and light particles. Curves: CDW-EIS results for proton (blue thin solid-line), antiproton (red dotted-line), electron (black thick solid-line) and positron (orange dashed-line) impact. Symbols: details in the inset; the references are: for protons p+ [15], and for electrons e- [21, 23 - 26, 29, 30].

For electron impact, the direct quintuple ionization is negligible even for energies near the threshold. Present curves represent the PCI, they are single ionization of inner-shell that ends as quintuple due to an Auger processes in cascade. It is worth to mention that to describe the quintuple ionization of Kr even the ionization of the L-shell must be considered [68]. Moreover, the good performance obtained with the CDW-EIS for this Kr^{5+} final charge state is due to the description of the ionization of the very deep shells, both by heavy and light projectiles.

The Multiple Ionization of Xe

Finally, Xenon multiple ionization by |Z=1| projectiles is displayed in Figs. (**14-18**), from single to quintuple ionization. The results for multiple ionization of Xe in these figures show the scope of the model that combines CDW-EIS, IPM for multiple ionization, mass effects in the description of light projectiles and the PCI through the empirical branching ratios. The theoretical results for Xe show a very good description of the experimental measurements for the different $|Z|=1$ projectiles, meaning that the charge effects (positive *vs.* negative charges, repulsive or attractive potential) and the mass effects (difference in the impact energy, threshold, non-linear trajectories) are correctly included. On the other hand, it is reasonable that the IPM approximation (no electron-electron correlation, no changes in the target potential due to the loss of one or more electrons) works better for Xe with 54 electrons than for Ne, with only 10 electrons. The Xe potential is less affected by the loss of a few electrons, and the approximation of independent events without changes is more realistic.

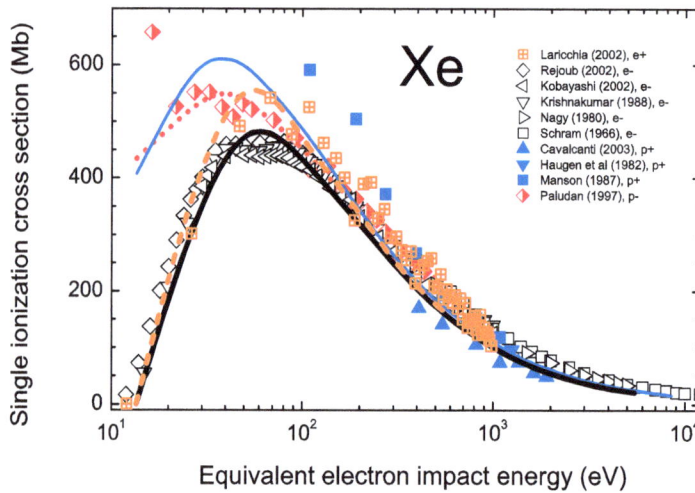

Fig. (14). Single ionization of Xe by |Z|=1projectiles as function of the impact energy, considering equal velocity heavy and light particles. Curves: CDW-EIS results for proton (blue thin solid-line), antiproton (red dotted-line), electron (black thick solid-line) and positron (orange dashed-line) impact. Symbols: details in the inset; the references are: for protons p+ [9, 14, 18], for electrons e- [21, 23 - 26, 29, 30] for positrons e+ [36], and for antiprotons p- [48].

Xe Single Ionization

In Fig. (**14**) the ionization of just one electron is displayed in linear scale in order to show in detail the measurements and the behavior of the cross sections around the maximum. The CDW-EIS predicts that proton and antiproton maximum cross sections are around 40 eV (in electron impact energy, it is $v \sim 1.7$ a.u.), while positron and electron curves are shifted to 55-60 eV. It is very interesting that the positron and antiproton highest cross sections are almost equal. The electron impact values are below the positron data around the maximum. For impact energies 40-80 eV, the single ionization by positron-impact is more effective (higher probabilities) than the electron-impact. This is charge effect. On the other hand, the comparison positron-proton and electron-antiproton gives a good idea of the mass effect.

The agreement with the experimental values reinforce these conclusions. The description of the antiproton measurements by Paludan *et al*. [48] is quite good in the whole energy range. The same for electron and positron impact. Clearly the positron measurements are above the electron ones, as predicted theoretically. Proton impact values at high energies converge to the others. The values by Manson *et al*.. [18] for protons at the three lower energies are clearly above the curve. The comparison of them with the single ionization by antiproton, which is actually pure ionization (no possible capture), suggests that electron capture may be present in these data. Measurements for Xe^+ production by proton impact at $1 \le v \le 4$ would be useful to be compared with the similar antiproton values.

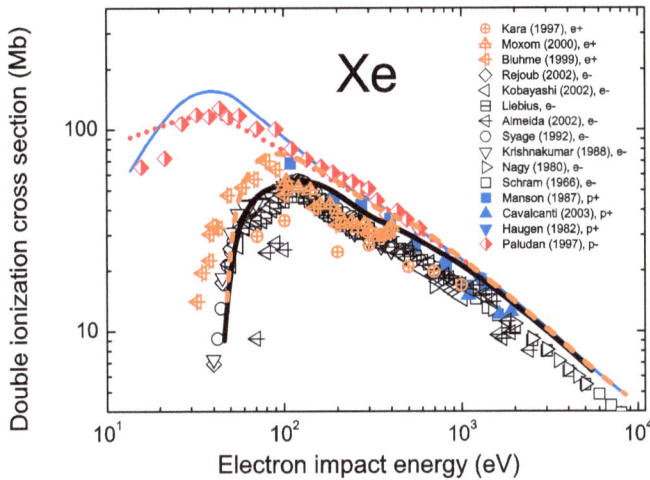

Fig. (15). Double ionization of Xe by |Z|=1 projectiles as function of the impact energy, considering equal velocity heavy and light particles. Curves: CDW-EIS results for proton (blue thin solid-line), antiproton (red dotted-line), electron (black thick solid-line) and positron (orange dashed-line) impact. Symbols: details in the inset; the references are: for protons p+ [9, 14, 18], for electrons e- [21, 23 - 26, 29, 30] for positrons e+ [39, 42, 43], and for antiprotons p- [48].

Xe Double Ionization

The agreement between the theoretical and the experimental values in Fig. (**15**) is amazing. These cross sections are not so small, *i.e.* around 10^{-15} - 10^{-16} cm². The experimental values for double ionization of Xe by antiprotons are very interesting because they cover the energy range around the maximum of the cross section. Instead, there is a lack of proton impact data in this energy region (velocity $v < 3$ a.u.). For positron impact, there are three groups of independent data with rather good agreement with the CDW-EIS results. It may be noted that the low energy measurements by Bluhme *et al*. [48] suggest a different energy threshold for electron and positron impact. This is not shown by the theory [56, 53] included here. However, it is worth to remember here that the CDW-EIS is not expected to be valid near the energy threshold. So we cannot state a conclusion about this positron-electron difference.

Above 1 keV all the double ionization cross sections converge. These values are highly influenced by the single ionization of a subvalence electron followed by decay and emission of a second electron. This PCI is important for energies above 200 eV and represents 80% of the double ionization cross section at 1 keV.

Xe Triple Ionization

The triple ionization of Xe, theory and compilation of available data, is presented in Fig. (**16**). The comparison shows again the validity of the description. The antiproton, proton and electron impact data are in reasonable agreement with the theoretical curves. For triple ionization, the threshold for light projectile ionization is very sharp and pushes down the cross sections. Theoretically, the cross sections are similar for electron and positron impact, with the two humps and a small minimum at 200 eV. At high energies, the positron curve converges to proton one at 400 eV, while the electron curve does above 1 keV. This high energy triple ionization is mostly PCI, as indicated in the figure. The curve obtained just considering the direct triple ionization (without PCI) is far from describing the data.

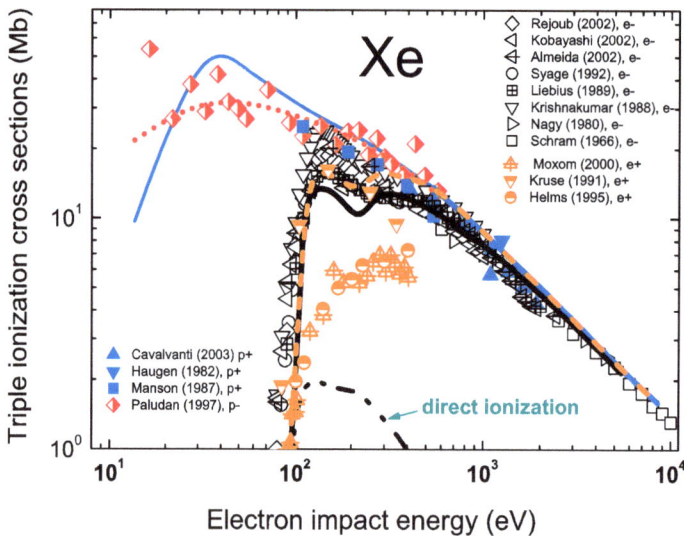

Fig. (16). Triple ionization of Xe by |Z|=1 projectiles as function of the impact energy, considering equal velocity heavy and particles. Curves: CDW-EIS results for proton (blue thin solid-line), antiproton (red dotted-line), electron (black thick solid-line) and positron (orange dashed-line) impact. Symbols: details in the inset; the references are: for protons p+ [9, 14, 18], for electrons e- [21, 23 - 26, 29, 30] for positrons e+ [43 - 45] and for antiprotons p- [48].

It can be noted in figure 16 that there are differences among the three sets of positron data. The measurements by Kruse *et al.* [51] agree with the expected values around the maximum. However, the more recent measurements by Moxom *et al.* [49] and Helms *et al* [50] are below the curves and the electron-impact data. Again we can draw attention to the lack of proton impact measurements at intermediate and low energies.

Xe Quadruple Ionization

In the Fig. (17) the quadruple ionization of Xe is displayed, including the experimental data available and the CDW-EIS results for the four $|Z|=1$ projectiles. The abundance of electron impact data can be noted, while there is only one set for positron impact [49], and only one set for proton impact [9]. No antiproton values are available in the literature, and proton data is available only in the energy region where they converge to electron-impact ones. The lack of measurements for the heavy projectiles around the maximum of the cross section does not allow to evaluate the theoretical predictions for proton and antiprotons. On the other hand, the positron measurements by Moxom *et al.* [49] are a factor 2 below the theoretical values.

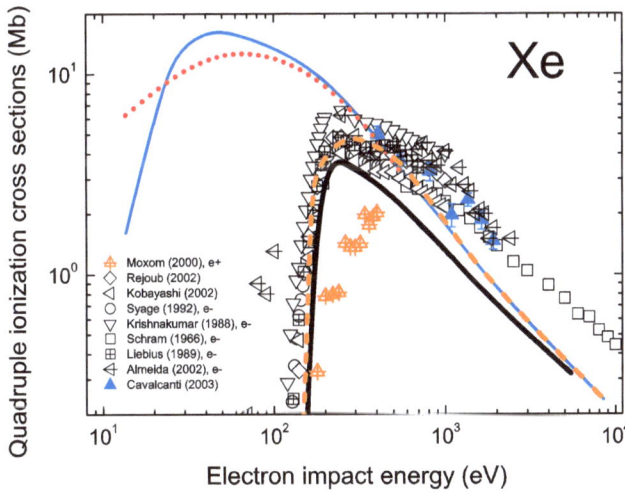

Fig. (17). Quadruple ionization of Xe by $|Z|=1$ projectiles as function of the impact energy, considering equal velocity heavy and light particles. Curves: CDW-EIS results for proton (blue thin solid-line), antiproton (red dotted-line), electron (black thick solid-line) and positron (orange dashed-line) impact. Symbols: details in the inset; the references are: for protons p+ [9], for electrons e- [21, 23 - 26, 29, 30] and for positrons e- [43].

The agreement with the electron impact data below 300 eV is fine considering they are quadruple-ionization cross sections. On the contrary, the theoretical high-energy values are too small. Concerning this undervalue, the CDW-EIS results in [53] include very deep shells, so the branching ratios of PCI may be the reason. However, it is fair to remark that at 5 keV the direct quadruple ionization cross sections are 4 orders of magnitude below the experimental values. Thus, the theoretical values including PCI, even a factor two below the data at 5 keV, are a rather good description of the electron impact values.

Xe Quintuple Ionization

Finally, for quintuple ionization of Xe in Fig. (**18**), only electron and proton impact data are available in the literature. For electron impact, the theoretical curve is in good agreement with the experimental measurements, considering the difficulties of such low values. The high energy proton measurements by Cavalcanti *et al.* [9] also agree with the electron data, except with those by Almeida *et al.* [29] that are larger than the rest. The theoretical curves for proton and antiproton impact are quite similar around the maximum. But there is no experimental data to confirm this.

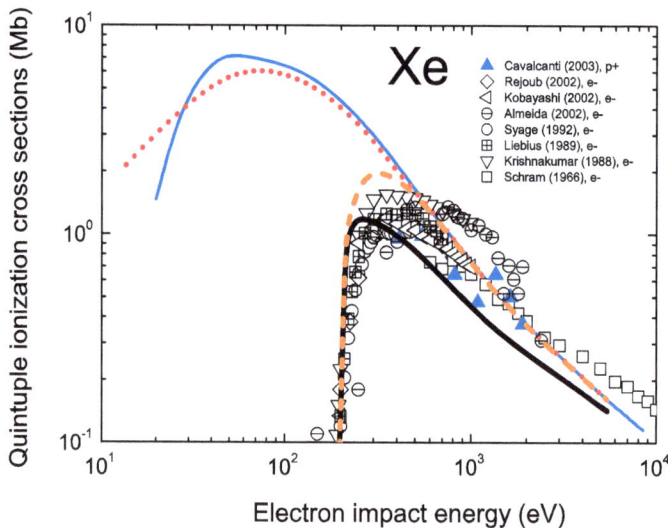

Fig. (18). Quintuple ionization of Xe by |Z|=1projectiles as function of the impact energy, considering equal velocity heavy and light particles. Curves: CDW-EIS results for proton (blue thin solid-line), antiproton (red dotted-line), electron (black thick solid-line) and positron (orange dashed-line) impact. Symbols: details in the inset; the references are: for protons p+ [9], and for electrons e- [21, 23, 24, 25, 26, 29, 30].

There are some characteristics to remark about the CDW-EIS results for quintuple ionization displayed in figure 18: the first one is that at low impact energies, the energy threshold for electron and positron impact is nicely described. The inclusion of the mean energy transferred in the threshold for multiple ionization proved to be important [56]. For example, while the outer 5s and up electrons of Xe have binding energies around 13 eV, the threshold for quintuple ionization is 178 eV [53], much greater than 5×13 eV. The second point to emphasize is that the Auger-like processes leading to quintuple ionization are by far the main contribution for light projectiles in the whole energy range, even near the threshold. The single ionization of the Xe 4s, 3d, and 3p electrons are the most important contributions to final quintuple ionization. This is clear in view of the experimental branching ratios for Xe in table 4 of [13]. At high energies, again the convergence of positron values to proton and antiproton takes place around 400 eV, while electron impact values do so only above 6 keV.

CONCLUDING REMARKS AND FUTURE PROSPECTS

In this chapter the particle and antiparticle multiple ionization of Ne, Ar, Kr and Xe was analyzed considering heavy and light projectiles. This was done based on the CDW-EIS results and a vast compilation of the experimental data available in the literature for the sixteen collisional systems, four projectiles and four targets.

The theoretical formalism combines the independent particle model, the CDW-EIS ionization probabilities for heavy projectiles, the changes in this approximation to consider light-particle collisions (difference in the impact energy, threshold, non-linear trajectories), and the inclusion of the post-collisional contribution to the final charge state due to Auger-like processes.

The multiple ionization of Kr and Xe shows the scope of the model: good description for the different $|Z|=1$ projectiles, meaning that the charge and mass effects are correctly included. The improvement of the theoretical results with the target nuclear charge (or with the number of bound electrons) is evident. The employment of the independent particle model to obtain multiple ionization probabilities does not consider changes in the target potential due to the loss of one or more electrons. It is reasonable that this approximation works better for atoms such as Kr or Xe than for Ne.

The correction of the CDW-EIS approximation to describe light particles shows the tendency of the experimental data. On the other hand, the energy thresholds calculated separately and imposed within the multinomial expansion, proved to describe rather well the experimental values for electron and positron impact. Although the threshold itself is rather well described, the CDW-EIS approximation is not expected to be valid for low energies.

The highly-charged ion production is dominated by the post-collisional electron emission that follows the inner-shell ionization (rearrangement and/or Auger processes). The CDW-EIS results are good even for quadruple and quintuple ionization cross sections. These calculations are sensitive to the good description of the deep shell ionization. They also rely on the empirical branching ratios of post-collisional ionization. We concluded that in multiple ionization by light particles, the post-collisional contribution is the main ionization channel in the whole energy range, even close to the threshold.

Future prospects can be separated in two lines. Within the theoretical work, the case of Ne is one of the limitations of the model. Not only related to the independent electron approximation, but also to the inclusion of the shake-off. Future progress in this regard is expected. On the other hand, the CDW-EIS has already been tested for sextuple ionization of Kr and Xe by electron impact with very good agreement with the experimental data. The extension to higher ionization orders (final charge state $+q \geq 7$ requires high computational effort but is possible.

Within the experimental research, the extensive compilation of data and the comparison of electron, positron, proton and antiproton values have been very enlightening. Unfortunately, the thorough particle-antiparticle comparison could be made up to double ionization. This is an interesting point to consider for future research. There are no positron impact experiments for higher levels of multiple ionization of rare gases. The exception is Xe. The antiproton data available goes up to triple ionization. It is interesting to note that, though the requirements and difficulties of the antiproton experiments (high energy physics facility), there are more measurements for ionization of Xe by antiprotons than by protons.

Present review also discussed the normalization of antiparticle relative measurements to the high energy electron ionization cross sections. We found that at high energies, the ionization cross sections by antiparticles impact converge to proton impact values at lower energies rather than to electron impact ones. Recent techniques for electron ionization measurements introduce very low relative errors, and make them quite interesting and reliable for the normalization purposes. The point to consider is that from which impact energy positron and electron, or antiproton and electron ionization cross sections, are equal.

CONSENT FOR PUBLICATION

Not applicable.

CONFLICT OF INTEREST

The author (editor) declares no conflict of interest, financial or otherwise.

ACKNOWLEDGMENTS

This work was partially supported by the following institutions from Argentina: the Consejo Nacional de Investigaciones Científicas y Técnicas, the Agencia Nacional de Promoción Científica y Tecnológica, and Universidad de Buenos Aires. The author thanks Jorge Miraglia for useful discussions on the CDW-EIS results.

BIBLIOGRAPHY

[1] J.H. McGuire, *Positron (Electron)-Gas Scattering.*, W.E. Kauppila, T.S. Stein, J.M. Wadehra, Eds., World Scientific: Singapore, 1986, pp. 222-231.

[2] D R Schultz, R E Olson, and C Reinhold, *Phys. B: At. Mol. Opt. Phys.,* vol. 24, pp. 521-558, 1991.

[3] H. Knudsen, and J.F. Reading, *Phys. Rep.,* vol. 212, pp. 107-222, 1992.
[http://dx.doi.org/10.1016/0370-1573(92)90013-P]

[4] H. Knudsen, and J.F. Reading, *Phys. Rep.,* vol. 212, pp. 107-222, 1992.
[http://dx.doi.org/10.1016/0370-1573(92)90013-P]

[5] K. Paludan, G. Laricchia, P. Ashley, V. Kara, J. Moxom, H. Bluhme, H. Knudsen, U. Mikkelsen, S.P. Möller, U. Uggerhöj, and E. Morenzoni, *J. Phys. At. Mol. Opt. Phys.,* vol. 30, pp. L581-L587, 1997.
[http://dx.doi.org/10.1088/0953-4075/30/17/005]

[6] R.D. DuBois, *Phys. Rev. Lett.,* vol. 52, pp. 2348-2351, 1984.
[http://dx.doi.org/10.1103/PhysRevLett.52.2348]

[7] G. Laricchia, D.A. Cooke, A. Köver, and S.J. Brawley, "Experimental Aspects of Ionization Studies by Positron and Positronium Impact (Cambridge University Press)",

[8] E G Cavalcanti, G M Sigaud, and E C Montenegro, *J. Phys. B: At. Mol. Opt. Phys.,* vol. 35, pp. 3937-3944, 2002.

[9] E.G. Cavalcanti, G.M. Sigaud, E.C. Montenegro, and H. Schmidt-Bocking, *J. Phys. At. Mol. Opt. Phys.,* vol. 36, pp. 3087-3096, 2003.
[http://dx.doi.org/10.1088/0953-4075/36/14/311]

[10] T. Spranger, and T. Kirchner, *J. Phys. At. Mol. Opt. Phys.,* vol. 37, p. 4159, 2004.
[http://dx.doi.org/10.1088/0953-4075/37/20/010]

[11] M.E. Galassi, R.D. Rivarola, and P.D. Fainstein, *Phys. Rev. A,* vol. 75, p. 052708, 2007.
[http://dx.doi.org/10.1103/PhysRevA.75.052708]

[12] C.C. Montanari, E.C. Montenegro, and J.E. Miraglia, *J. Phys. At. Mol. Opt. Phys.,* vol. 43, p. 165201, 2010.
[http://dx.doi.org/10.1088/0953-4075/43/16/165201]

[13] C.C. Montanari, and J.E. Miraglia, *J. Phys. At. Mol. Opt. Phys.,* vol. 47, p. 105203, 2014.
[http://dx.doi.org/10.1088/0953-4075/47/10/105203]

[14] H K, L.H. Andersen, P. Hvelplund, and H. Knudsen, *Phys. Rev A.,* vol. 26, pp. 1962-1974, 1982.

[15] R.D. DuBois, L.H. Toburen, and M. Rudd, "E1984", *Phys. Rev. A,* vol. 29, pp. 70-76, .
[http://dx.doi.org/10.1103/PhysRevA.29.70]

[16] L H Andersen, P Hvelplund, H Knudsen, S P Möller, A H Sörensen, K Elsener, and K G Rensfelt, *Phys. Rev. A.,* vol. 36, pp. 3612-3629, 1987.

[17] R.D. DuBois, and S.T. Manson, *Phys. Rev. A,* vol. 35, pp. 2007-2025, 1987.
[http://dx.doi.org/10.1103/PhysRevA.35.2007]

[18]   S T Manson, and R D DuBois, *J. Physique.,* vol. 48, pp. C9 263-266, 1987.

[19]   A.D. González, and E.H. Pedersen, "Differential cross sections for the multiple ionization of Ne and Ar by protons", *Phys. Rev. A,* vol. 48, no. 5, pp. 3689-3698, 1993.
       [http://dx.doi.org/10.1103/PhysRevA.48.3689] [PMID: 9910038]

[20]   L. Sarkadi, P. Herczku, S.T.S. Kovacs, and A. Kover, *Phys. Rev. A,* vol. 87, p. 062705, 2013.
       [http://dx.doi.org/10.1103/PhysRevA.87.062705]

[21]   B.L. Schram, A.J.H. Boerboom, and J. Kistermaker, "Physica 32 185-196; Schram B L 1966 Physica 32 197-209; Schram B L, de Heer F J, Van der Wiel M J and Kistermaker J 1965 \ Physica 31 94; Adamczyk B, Boerboom A J H, Schram B L and Kistermaker J 1966", *J. Chem. Phys.,* vol. 44, pp. 4640-4642, 1966.

[22]   P. Nagy, A. Skutlartz, and V. Schmidt, *J. Phys. At. Mol. Opt. Phys.,* vol. 13, pp. 1249-1267, 1980.
       [http://dx.doi.org/10.1088/0022-3700/13/6/028]

[23]   J.A. Syage, "Electron-impact cross sections for multiple ionization of Kr and Xe", *Phys. Rev. A,* vol. 46, no. 9, pp. 5666-5679, 1992.
       [http://dx.doi.org/10.1103/PhysRevA.46.5666] [PMID: 9908816]

[24]   E. Krishnakumar, and S.K. Srivastava, *J. Phys. At. Mol. Opt. Phys.,* vol. 21, pp. 1055-1082, 1988.
       [http://dx.doi.org/10.1088/0953-4075/21/6/014]

[25]   R. Rejoub, B.G. Lindsay, and R.F. Stebbing, *Phys. Rev. A,* vol. 65, p. 042713, 2002.
       [http://dx.doi.org/10.1103/PhysRevA.65.042713]

[26]   A. Kobayashi, G. Fujiki, A. Okaji, and T. Masuoka, *J. Phys. At. Mol. Opt. Phys.,* vol. 35, pp. 2087-2103, 2002.
       [http://dx.doi.org/10.1088/0953-4075/35/9/307]

[27]   P. McCallion, M.B. Shah, and H.B. Gilbody, *J. Phys. At. Mol. Opt. Phys.,* vol. 25, pp. 1061-1071, 1992.
       [http://dx.doi.org/10.1088/0953-4075/25/5/017]

[28]   H.C. Straub, P. Renault, B.G. Lindsay, K.A. Smith, and R.F. Stebbings, "Absolute partial and total cross sections for electron-impact ionization of argon from threshold to 1000 eV", *Phys. Rev. A,* vol. 52, no. 2, pp. 1115-1124, 1995.
       [http://dx.doi.org/10.1103/PhysRevA.52.1115] [PMID: 9912350]

[29]   D.P. Almeida, A.C. Fontes, and C F L. Godinho, *J. Phys. At. Mol. Opt. Phys.,* vol. 28, pp. 3335-3345, 1995.
       [http://dx.doi.org/10.1088/0953-4075/28/15/022]

[30]   H. Liebius, J. Binder, H.R. Koslowwski, K. Wiesemann, and A. Huber, *J. Phys. At. Mol. Opt. Phys.,* vol. 22, pp. 83-97, 1989.
       [http://dx.doi.org/10.1088/0953-4075/22/1/011]

[31]   R.K. Singh, R. Hippler, and R. Shanker, *J. Phys. At. Mol. Opt. Phys.,* vol. 35, pp. 3243-3256, 2002.
       [http://dx.doi.org/10.1088/0953-4075/35/15/302]

[32]   H.R. Koslowski, J. Binder, B.A. Huber, and K. Wiesemann, *J. Phys. At. Mol. Opt. Phys.,* vol. 20, p. 5903, 1987.
       [http://dx.doi.org/10.1088/0022-3700/20/21/032]

[33]   H. Knudsen, L. Brun-Nielsen, M. Charlton, and M.R. Poulsen, *J. Phys. At. Mol. Opt. Phys.,* vol. 23, pp. 3955-3976, 1990.
       [http://dx.doi.org/10.1088/0953-4075/23/21/026]

[34]   F M Jacobsen, and N P Frandsen, "[3] H, Mikkelsen U and Schrader D M", *J. Phys. B: At. Mol. Opt. Phys.,* vol. 48, pp. 4691-4695, 1995.

[35]   S. Mori, and O. Sueoka, *J. Phys. At. Mol. Opt. Phys.,* vol. 27, pp. 4349-4364, 1994.

[http://dx.doi.org/10.1088/0953-4075/27/18/028]

[36]   G. Laricchia, P. Van Reeth, M. Szuinska, and J. Moxom, *J. Phys. At. Mol. Opt. Phys.,* vol. 35, pp. 2525-2540, 2002.
[http://dx.doi.org/10.1088/0953-4075/35/11/311]

[37]   J.P. Marler, J.P. Sullivan, and C.M. Surko, *Phys. Rev. A,* vol. 71, p. 022701, 2005.
[http://dx.doi.org/10.1103/PhysRevA.71.022701]

[38]   M. Charltoni, L. Brun-Nielsen, B.I. Deutch, P. Hvelplund, F.M. Jacobsen, H. Knudsen, G. Laricchiat, and M.R. Poulsen, *J. Phys. At. Mol. Opt. Phys.,* vol. 22, pp. 2779-2788, 1989.
[http://dx.doi.org/10.1088/0953-4075/22/17/016]

[39]   V. Kara, K. Paludan, J. Moxom, P. Ashley, and G. Laricchia, *J. Phys. At. Mol. Opt. Phys.,* vol. 30, pp. 3933-3949, 1997.
[http://dx.doi.org/10.1088/0953-4075/30/17/019]

[40]   J. Moxom, P. Ashley, and G. Laricchia, *Can. J. Phys.,* vol. 74, p. 367, 1996.
[http://dx.doi.org/10.1139/p96-053]

[41]   J Moxom, D. M. Schrader, G. Laricchia, and Jun Xu, *Phys. Rev. A.,* vol. 60, pp. 2940-2943, 1999.

[42]   H. Bluhme, H. Knudsen, J.P. Merrison, and K.A. Nielsen, *J. Phys. At. Mol. Opt. Phys.,* vol. 32, p. 5237, 1999.
[http://dx.doi.org/10.1088/0953-4075/32/22/302]

[43]   J. Moxom, *J. Phys. At. Mol. Opt. Phys.,* vol. 33, pp. L481-L485, 2000.
[http://dx.doi.org/10.1088/0953-4075/33/13/102]

[44]   S. Helms, U. Brinkmann, J. Deiwiks, R. Hippler, H. Schneider, D. Segers, and J. Paridaens, *J. Phys. B: At. Mol. Opt.,* vol. 28, p. 1095, 1995.
[http://dx.doi.org/10.1088/0953-4075/28/6/022]

[45]   G. Kruse, A. Quermann, W. Raith, G. Sinapius, and M. Weber, *J. Phys. At. Mol. Opt. Phys.,* vol. 24, p. L33, 1991.
[http://dx.doi.org/10.1088/0953-4075/24/2/003]

[46]   L H Andersen, P Hvelplund, H Knudsen, S P Möller, K Elsener, and K G Rensfelt, *Phys. Rev. Lett,* vol. 57, pp. 2147-2150, 1986.

[47]   L.H. Andersen, P. Hvelplund, H. Knudsen, S.P. Möller, J.O.P. Pedersen, S. Tang-Pedersen, E. Uggerhöj, K. Elsener, and E. Morenzoni, *Phys. Rev. A,* vol. 40, p. 7366, 1989.
[http://dx.doi.org/10.1103/PhysRevA.40.7366]

[48]   K. Paludan, H. Bluhme, H. Knudsen, U. Mikkelsen, S.P. Möller, E. Uggerhöj, and E. Morenzoni, *J. Phys. At. Mol. Opt. Phys.,* vol. 30, p. 3951, 1997.
[http://dx.doi.org/10.1088/0953-4075/30/17/020]

[49]   H Knudsen, *Phys. Rev. Lett ,* vol. 101, pp. 043201 107-222, 2008.

[50]   H. Knudsen, "Nucl. Instrum and Meth. in Phys", *Research Section B,* vol. 267, pp. 244-247, 2009.

[51]   T. Kirchner, and H. Knudsen, *J. Phys. At. Mol. Opt. Phys.,* vol. 44, p. 122001, 2011.
[http://dx.doi.org/10.1088/0953-4075/44/12/122001]

[52]   C.C. Montanari, and J.E. Miraglia, *J. Phys. At. Mol. Opt. Phys.,* vol. 45, p. 105201, 2012.
[http://dx.doi.org/10.1088/0953-4075/45/10/105201]

[53]   C.C. Montanari, and J.E. Miraglia, *J. Phys. At. Mol. Opt. Phys.,* vol. 48, p. 165203, 2015.
[http://dx.doi.org/10.1088/0953-4075/48/16/165203]

[54]   J.E. Miraglia, and M.S. Gravielle, *Phys. Rev. A,* vol. 78, p. 052705, 2008.
[http://dx.doi.org/10.1103/PhysRevA.78.052705]

[55]   P.D. Fainstein, V.H. Ponce, and R.D. Rivarola, *J. Phys. At. Mol. Opt. Phys.,* vol. 21, p. 287, 1988.

[http://dx.doi.org/10.1088/0953-4075/21/2/013]

[56]   C.C. Montanari, and J.E. Miraglia, *J. Phys. Conf. Ser.,* vol. 583, p. 012018, 2015.
        [http://dx.doi.org/10.1088/1742-6596/583/1/012018]

[57]   A C F Santos, and W S Melo, *Phys Rev A,* vol. 63, p. 062717, 2001.

[58]   M.M. Sant'Anna, H. Luna, A.C.F. Santos, C. McGrath, M.B. Shah, E.G. Cavalcanti, G.M. Sigaud, and
        E.C. Montenegro, *Phys. Rev. A,* vol. 68, p. 042707, 2003.
        [http://dx.doi.org/10.1103/PhysRevA.68.042707]

[59]   W Wolff, H Luna, A C F Santos, and E C Montenegro, *Phys. Rev. A,* vol. 84, p. 042704, 2011.

[60]   C.C. Montanari, J.E. Miraglia, W. Wolff, H. Luna, A.C.F. Santos, and E.C. Montenegro, *J. Phys.
        Conf. Ser.,* vol. 388, p. 012036, 2012.
        [http://dx.doi.org/10.1088/1742-6596/388/1/012036]

[61]   T. Kirchner, M. Horbatsch, and H.J. Lüdde, "Phys. Rev. A 64, 012711; 2002", *Phys. Rev. A,* vol. 66, p.
        052719, 2001.
        [http://dx.doi.org/10.1103/PhysRevA.66.052719]

[62]   T. Kirchner, A.C.F. Santos, H. Luna, M.M. Sant'Anna, W.S. Melo, G.M. Sigaud, and E.C.
        Montenegro, *Phys. Rev. A,* vol. 72, p. 012707, 2005.
        [http://dx.doi.org/10.1103/PhysRevA.72.012707]

[63]   G. Schenk, and T. Kirchner, *J. Phys. At. Mol. Opt. Phys.,* vol. 42, p. 205202, 2009.
        [http://dx.doi.org/10.1088/0953-4075/42/20/205202]

[64]   G. Schenk, M. Horbatsch, and T. Kirchner, *Phys. Rev. A,* vol. 88, p. 012712, 2013.
        [http://dx.doi.org/10.1103/PhysRevA.88.012712]

[65]   G. Schenk, and T. Kirchner, *Phys. Rev. A,* vol. 91, p. 052712, 2015.
        [http://dx.doi.org/10.1103/PhysRevA.91.052712]

[66]   C.A. Tachino, M.E. Galassi, and R.D. Rivarola, *Phys. Rev. A,* vol. 77, p. 032714, 2008.
        [http://dx.doi.org/10.1103/PhysRevA.77.032714]

[67]   C.A. Tachino, M.E. Galassi, and R.D. Rivarola, *Phys. Rev. A,* vol. 80, p. 014701, 2009.
        [http://dx.doi.org/10.1103/PhysRevA.80.014701]

[68]   A.C. Tavares, C.C. Montanari, J.E. Miraglia, and G.M. Sigaud, *J. Phys. At. Mol. Opt. Phys.,* vol. 47, p.
        045201, 2014.
        [http://dx.doi.org/10.1088/0953-4075/47/4/045201]

[69]   C.D. Archubi, C.C. Montanari, and J.E. Miraglia, *J. Phys. At. Mol. Opt. Phys.,* vol. 40, p. 943, 2007.
        [http://dx.doi.org/10.1088/0953-4075/40/5/010]

[70]   D. Rapp, and P. Englander-Golden, *J. Chem. Phys.,* vol. 43, p. 1464, 1965.
        [http://dx.doi.org/10.1063/1.1696957]

[71]   A.A. Sorokin, L.A. Shmaenok, S.V. Bobashev, B. Möbus, and G. Ulm, *Phys. Rev. A,* vol. 58, p. 2900,
        1998.
        [http://dx.doi.org/10.1103/PhysRevA.58.2900]

[72]   M.E. Rudd, Y-K. Kim, D.H. Madison, and J.W. Gallagher, *Rev. Mod. Phys.,* vol. 57, pp. 965-994,
        1985.
        [http://dx.doi.org/10.1103/RevModPhys.57.965]

CHAPTER 6

Inner Shell Ionization and Excitation of CCl_4 and its Relation to Electron Scattering

A.C.F. Santos[1], W.C. Stolte[2,*], G.G.B. Souza[3], M.M. Sant'Anna[1] and K.T. Leung[4]

[1] *Instituto de Fisica, Universidade Federal do Rio de Janeiro,Caixa Postal 68525, 21941-972, Rio de Janeiro, RJ, Brazil*

[2] *Advanced Light Source, Lawrence Berkeley National Laboratory, Berkeley, CA, 94720, USA*

[3] *Instituto de Quimica, Universidade Federal do Rio de Janeiro, Rio de Janeiro, RJ, Brazil, 21949-900*

[4] *Department of Chemistry, University of Waterloo, Canada, N2L 3G1*

Abstract: The fragmentation of the tetrachloromethane molecule, following core-shell photoexcitation and photoionization in the neighborhood of the chlorine *K*-edge has been studied by using time-of-flight mass spectroscopy and monochromatic synchrotron radiation. Branching ratios for ionic dissociation were derived for all detected ions, which are informative of the decay dynamics and photofragmentation patterns of the core-excited species. In addition, the absorption yield has been measured, with a new assignment of the spectral features. The structure that appears above the Cl 1s ionization potential in the photoionization spectrum, has been ascribed in terms of the existing connection with electron-CCl_4 scattering through experimental data and calculations for low-energy electron-molecule cross sections. In addition, the production of the doubly ionized Cl^{2+} as a function of the photon energy has been analysed in terms of a simple and appealing physical picture, the half-collision model.

PACS numbers: 33.80.-b, 33.80.Eh

Keywords: Photofragmentation, Ion branching ratios, Deep core, Ionization, Photoionization, CCl_4, Tetrachloromethane, Synchrotron radiation, Chlorine *K*-edge, Cl 1s, Photoabsorption, X-rays.

I. INTRODUCTION

Absorption of a soft x-ray photon (0.12-5 keV) by a target may lead to the transfer of an electron from a deep core to an empty orbital or to its ionization. A core-hole can be created through the absorption of an incoming x-ray photon with

---

* **Corresponding author A.C.F. Santos:** Instituto de Fisica, Universidade Federal do Rio de Janeiro,Caixa Postal 68525, 21941-972, Rio de Janeiro, RJ, Brazil; Tel: +55 21 3938-7947; Fax: +55 21 3938-7368; E-mail: toni@if.ufrj.br

Antônio Carlos Fontes dos Santos (Ed.)

energy matched to the binding energy of a core electron or even larger. This absorption process gives rise to a core excitation to a bound state or to the continuum where, in this case, the excited electron will become a free particle, *i.e.*, a photoelectron. After excitation or ionization, the absorbing molecule is found in a highly excited state due to the creation of the core hole. Shortly after the atom relaxes *via* the core hole decay. Thus, the photoexcited or photoionized molecule with holes in its deep core-shell of width ΔE is substantially unstable and short-lived with decay times of about 1 fs or few hundreds of attoseconds, given by the well-known relation:

$$\tau = \frac{\hbar}{\Delta E} \tag{1}$$

These inner-shell excited states decay by ejecting a photon (fluorescence), one or more electrons (Auger and/or Coster-Kronig cascades), giving rise to multiply charged moieties, which fragment in a very short time-scale [1]. Fig. (**1**) shows a sketch of an inner-shell decay process. An inner shell (K or L) hole is filled by an electron from an outer shell or another inner-shell. The energy difference between the shells can be transferred to another electron, which is then ionized from the target. Depending on the origin of the electron filling the vacancy, the emitted electrons are called Auger electrons (same shell) or Coster-Kronig (different shells) electrons.

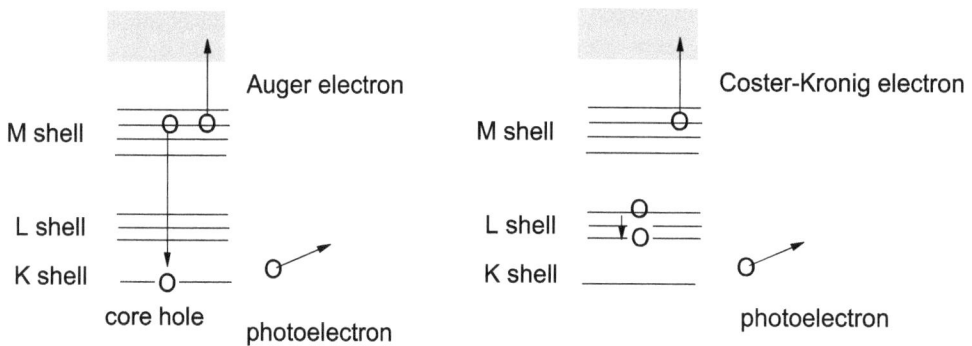

Fig. (1). Sketch of an inner-shell decay process. An inner shell (K or L) hole is filled by an electron from an outer shell or another inner-shell. The energy difference between the shells can be transferred to another electron, which is then ionized from the target. Depending on the origin of the electron filling the vacancy, the emitted electrons are called Auger electrons or Coster-Kronig electrons.

For photon energies where the Compton scattering can be neglected, one photon can remove no more than one electron from the molecule. Notwithstanding, eventually two electrons can be ejected even though the incident photon interacts with merely one electron in the molecule. This is possible due to electron-electron interaction *via* the so-called knock out and shake-off mechanisms [2, 3]. Thus, the ejection of two or more electrons from the target is a suitable probe for studying electron correlations. For atomic targets and for valence and shallow core excitation, the ejection of two or more electrons has small probabilities compared to the single-photoionization process and is in general, a few percent [4, 5]. The same is not true for deep core excitation, where multiple charged ions can be formed with significant probabilities. Additionally, two or more electrons can be ejected *via* Auger (or Coster-Kronig) decay after producing a core hole. It can also be more complicated, involving not only a core electron but also valence electrons [6]. As the photon energy increases, the ratio of multiply to singly ionized ions usually also increases up to a maximum from where it gradually decreases towards the so-called shake-off (SO) limit. For excess energies no more that a few 100 eV, the double ionization process can be ascribed as an internal collision (the so-called knock out (KO) or TS-1 mechanism). The photoelectron can interact with the second electron during its way out of the target producing two ejected electrons. Thus the double photoionization process should be similar to the electron impact by the photoelectron that leaves the atom at a low speed. Thus, the ratio of double to single ionization at low excess energies above the ionization potential, which is the probability for ionization of the second electron, should be proportional to the single electron-impact ionization cross section of the singly charged moiety. Fig. (**2**) shows a sketch of the shake-off (SO) and knock-out (KO or TS-1) mechanisms.

Another scope of this paper is to discuss the shape resonance (SR) observed above the Cl 1s threshold in the CCl_4 molecule. SR is a continuum resonant process that can be observed in small molecules [7]. It ordinarily appears, within a few eV above an ionization potential, as a broad continuum structure in the photoabsorption and photoionization spectra. Molecular SR can be attributed to the trapping of the photoelectron by a potential barrier through which the electron sooner or later tunnels and is ejected [7]. In this qualitative sketch, the details of the molecular potential take a paramount role, which shape supports a potential barrier, due to the roles of the attractive and repulsive parts of the potential, due to the Coulombic and centrifugal forces, respectively. This interplay between attractive and repulsive forces controls the shape of the potential. Another picture connects to the chemical features of shape resonances, associating them to empty molecular orbitals close to the continuum. There are important parallels between the both pictures which have been pointed out [7]. The connection between electron-molecule scattering and molecular photoionization, with emphasis on

SRs, has been suggested by Dehmer and Dill [8]. They observed that although electron-molecule scattering and molecular photoionization have different electron numbers, the short-range nature of shape resonances preserves the similarity, even having the long-range part of the potential substantially altered. In this paper, we present a model where the photo-excited core-shell electron behaves as a projectile, and the residual ionic molecule as a target in an electron attachment collision.

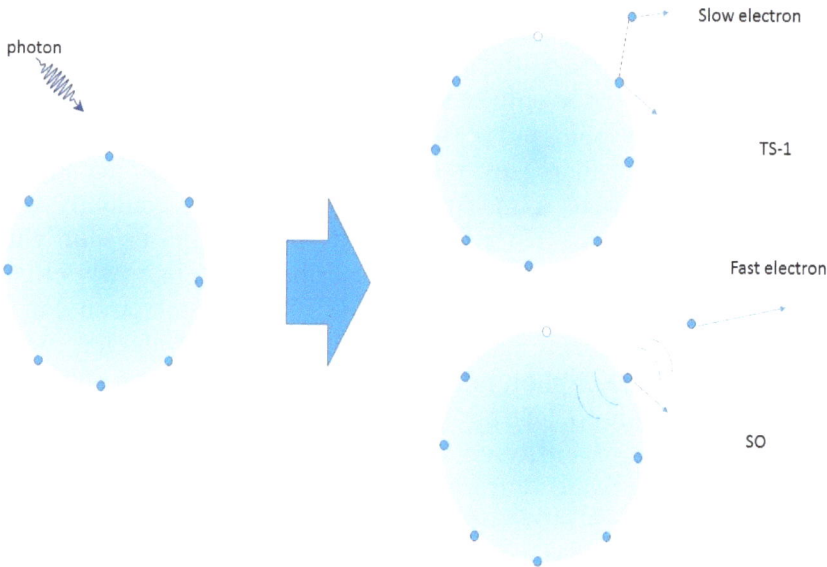

Fig. (2). Sketch of the shake-off (SO) and knock-out (KO or TS-1) mechanisms.

Carbon tetrachloride, tetrachloromethane or CCl_4 is a tetrahedral molecule in its ground state and has been subject of study for a long time due to the its technological applications as etchant gas in microelectronics [9, 10]. The previously measured spectra for Inner shell photoabsorption of the CCl4 molecule [11 - 14] were restricted to the region of Cl $2p$. Zhang *et. al.* [15] reported absolute dipole differential oscillator strengths for Cl K-shell spectra from high-resolution electron energy loss studies of CCl_4 among other molecules. CCl_4 is a closed-shell molecule belonging, in its ground state, to the symmetry point T_d. Fig. (**3**) shows the energy levels of the CCl_4 molecule in an independent particle approximation. The $(7a_1)^0$ and $(8t_2)^0$ are the lowest unoccupied molecular orbitals (LUMO) [15].

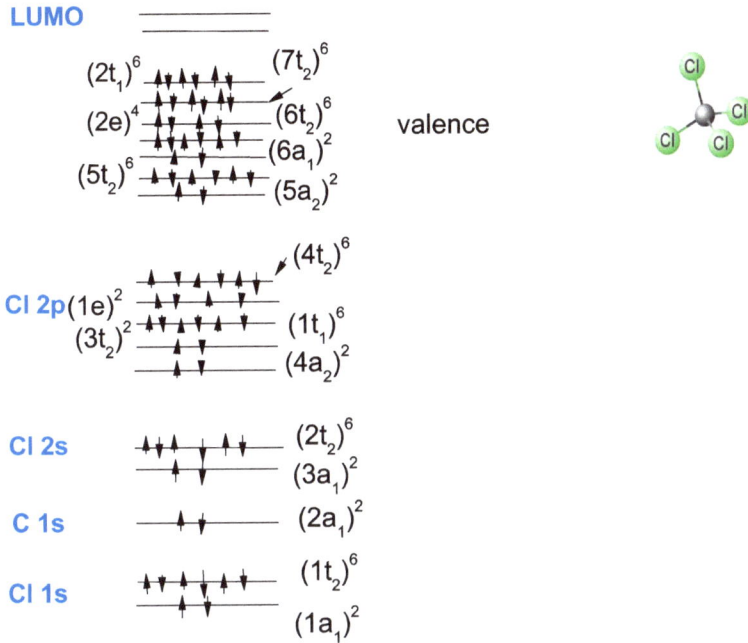

Fig. (3). Energy levels of the CCl_4 molecule.

II. EXPERIMENTAL TECHNIQUE

Measurements were performed at the Advanced Light Source (Lawrence Berkeley National Laboratory, Berkeley, CA, USA) on bending magnet beamline 9.3.1, which includes a Si(111) double-crystal monochromator with a resolution of approximately 0.45 eV [17], and nearly 100% linearly polarized x-rays near the Cl 1s-edge... Calibration was made by comparison to an absorption measurement of CF_3Cl, and using its Cl $(1s)^{-1} \rightarrow 11a_1$ transition at 2823.5 eV [16].

The presented branching ratios were measured with a time-of-flight (TOF) mass spectrometer, which was used in the space-focusing mode [18], under conditions similar to our previous experimental setup [19]. The total distance traveled by the ions was about 8 cm. The first stage of the electric field consists of a plate-grid system with the light beam passing in its middle with a 2200 V/cm electric field. The gas inlet needle was kept at ground potential, and the drift tube was kept at 3700 V. The heavier ions required several 328 ns time periods to arrive at the detector, whereas, hydrogen arrived in approximately 210 ns. Additionally, to keep a constant efficiency, the ions were required to have sufficient energy such that they strike the front surface of the microchannel plate (MCP) with an energy between 4 and 5 keV, thus the MCP's had a -4500 V voltage applied to the front surface. The output pulses from the MCP were fed into conventional electronic counting equipment, with constant fraction discriminators (Ortec model 9327) being used to remove any electronic noise. The discrimination level was kept as low as possible, less than 15 mV. The levels were verified by comparing the branching ratios for the fragment ions from an argon target, on the Ar $1s \rightarrow 4p$ resonance [Levin PRA **65** (1990) 988] while varying the discriminator setting. The target chamber base pressure was in the 10^{-9} Torr range. In order to minimize the contribution from false coincidences the pressure during the experiment was kept in the mid 10^{-6} Torr range. The ion signal was used to start the time-t--amplitude converter (TAC), and a signal from the synchrotron ring provided a reliable stop pulse for the TAC.

The presented absorption spectrum, I/I_0, (see Fig. **1**) was measured using a static gas cell, containing 35 Torr of CCl_4. The cell was closed on both sides with 500 μm thick Si_3N_4 windows of 2 mm diameter, with a path length of 1.56 cm. The gas cell was preceded in the beamline by an aluminized Mylar foil which was used to measure I_0. The absorption signal I was measured with a Si-photodiode (International Radiation Detectors, Inc., model AXUV100), and the currents from both the aluminized Mylar and the photodiode were monitored with a pair of calibrated Keithley model 6517A electrometers.

The tetrachloromethane was commercially obtained as a 99.9% pure liquid from Sigma Aldrich. The sample was purged of contaminants by performing a minimum of three freeze/thaw cycles while pumping out the excess vapors from the frozen sample with a oil free mechanical pump, and finally introduced into the chamber after thawing by using its vapor pressure at room temperature.

III. RESULTS

III.1. Photoabsorption Spectra

Fig. (**4**) shows the total ion yield spectrum of tetrachloromethane near the Cl *K*-edge. The total ion-yield spectrum mimics the photoabsorption in the present energy range due to the small probability for relaxation though fluorescence. The main peak in the Fig. (**1**) (2823.4 eV) is the result of Cl $1s$ resonant excitation $(1t_2)^6(1a_1)^2 \rightarrow 7a_1$.

There is a clear structure above the 1s threshold (see Fig. **4**). We attribute its origin to a shape resonance. This assignment, however, is not straightforward. In a study of L-shell photoionization of CCl_4, Lu *et al.* attributed peaks measured about 1 eV above the Cl $2p_{1/2}$ and Cl $2p_{3/2}$ thresholds to delay onset and shake-up effects [10, 20]. Only a broader structure, about 9 eV above threshold was assigned by them to a shape resonance. In the absence of theoretical calculations

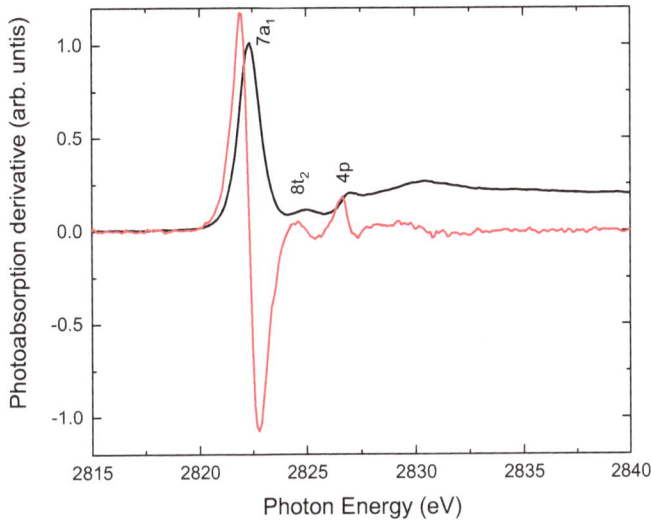

Fig. (4). Photoabsorption (black line) and its derivative (red line) of the CCl_4 molecule around the Cl 1s edge.

for the Cl 1s photoionization, we seek information using a parallel with experimental data and calculations for low-energy electron-molecule cross sections. This connection, with emphasis on shape resonances, has been pointed by Dehmer and Dill [8]. They note that electron-molecule scattering ($e^- + M$) and molecular photoionization ($hv+M$) systems have different electron numbers. Notwithstanding, they argue that the short-range nature of shape resonances preserves the similarity, even having the long-range part of the potential substantially altered. Loosely speaking, we picture in the following comparison the photo-excited core-shell electron (in $hv + M$) behaving as a projectile, and the residual ionic molecule as a target in an electron attachment collision (a hypothetical $e^- + M(1s^{-1})$).

Braun *et al.* have measured high-resolution electron attachment cross sections for a CCl4 molecular target [21]. In general, temporary negative ions (TNI), are formed and later decay in several channels. Cross sections presented for the Cl_2^- channel show a peak 1.3 eV above threshold with FWHM of 0.6 eV. Jones measured absolute total cross sections for e- + CCl_4 scattering at low electron energies [22]. In order to interpret the measured structures (see Fig. **2**) Jones makes use of calculations using the MS-Xα method which provide 4 peaks: two are interpreted as nonvalence shape resonances of t_2 and e symmetries at 1.74 eV and 6.3 eV, respectively.

The other two structures (less intense), are a_1 and e resonances found at 9.4 eV and 13.3 eV. These assignments are not consensual. Burrow *et al.* measured the low-energy peak at 0.94 eV in an electron transmission spectrum [23] and attributed its origin to a capture of the incident electron in a triply degenerate carbon chlorine antibonding orbital (C-Cl σ^*) of t_2 symmetry. Curik *et al.* calculated elastic integral cross sections and discussed the contributions of the A_1, T_2, and E symmetries of the T_d group [24]. In their calculations, the lowest energy peak has a dominant A_1 contribution and the second peak a dominant T_2 contribution. A contribution of E symmetry shows two additional maxima. Their results are shown in Fig. (**5**) together with the experimental data of Jones *et al.* [22] for total electron scattering cross sections and the present results for photon absorption, as a function of the energy of the projectile minus the threshold energy. Calculations in the static-exchange approximation of Natalense *et al.* also show three main structures in the A_1, T_2, and E symmetries of the T_d group [25]. Recently, Moreira *et al.* presented calculations based on the Schwinger multichannel method with pseudopotentials and reported two shape resonances at 0.75 eV and 8 eV belonging to the T_2 and E symmetries [26].

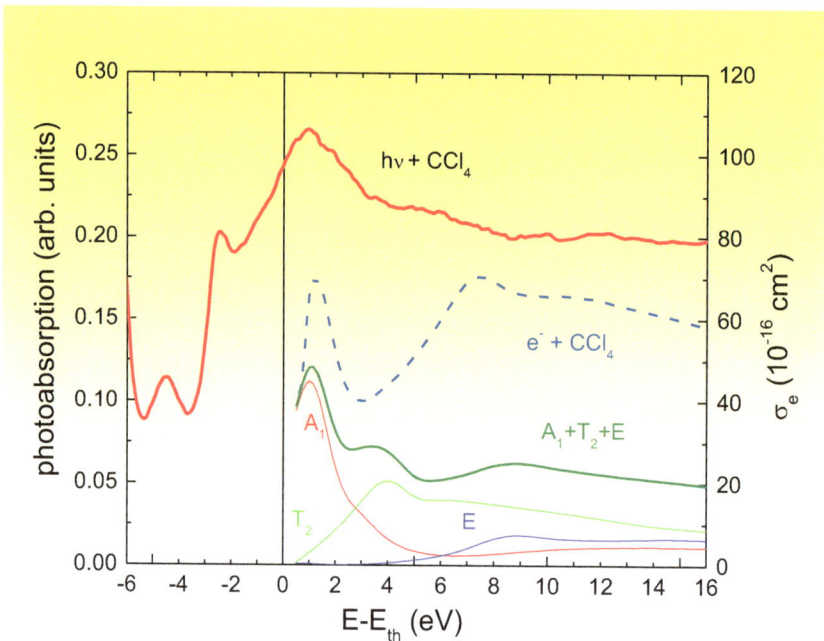

Fig. (5). Photoabsorption of the CCl_4 molecule around the Cl 1s edge (photon + CCl_4) as a function of the photon energy in comparison to electron scattering on CCl_4.

In Fig. (**6**) we show our data for photon absorption and the results for a least square fit of Lorentzians. The two first peaks are attributed to the two shape resonances discussed in the e- + CCl_4 studies mentioned above. With respect to the 1s ionization potential, the most intense peak is located at 1.0eV and has a linewidth of 3.6 eV. The second most intense peak is located at 5.5 eV and has a linewidth of 3.4 eV. The two peaks marked with "*" at Fig. (**3**) may correspond to the higher energy structures mentioned by Jones *et al.* [22] at 9.4 eV and 13.3 eV. However, the background in our measurements do not allow unequivocal assignments of these peaks.

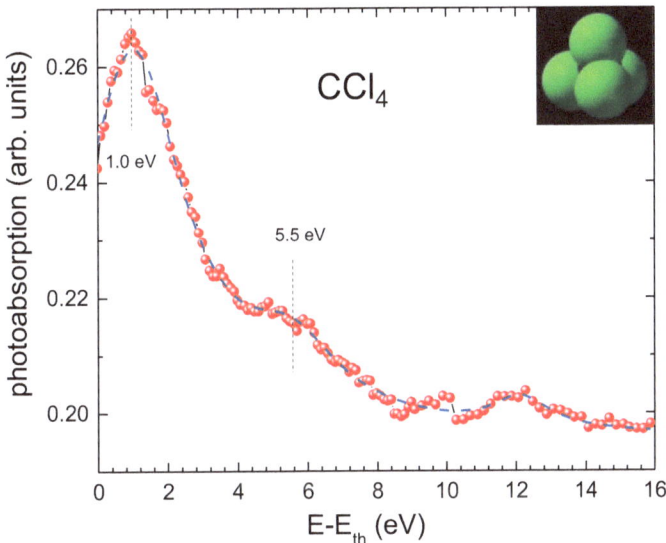

Fig. (6). Data for photon absorption and the results for a least square fit of Lorentzians. The two first peaks are attributed to the two shape resonances discussed in the e- + CCl_4 studies (see text for details).

III.2. Ion branching Ratios

Ion-yield spectra were taken in the photon energy range of 2802–2872 eV. In Fig. (**7a**) we present a overview of the branching ratios for the most abundant fragments of CCl_4 as a function of the photon energy around the Cl 1s edge. The spectra is dominated by the Cl^+ fragment (see Fig. **7b**), as expected (~75% below the Cl edge and ~ 52% above the edge). Its relative intensity shown a local minimum at the $(1t_2)^6(1a_1)^2 \rightarrow 7a_1$ resonance, a weak local maximum around the $(1t_2)^6(1a_1)^2 \rightarrow 8t_2$ and $(1t_2)^6(1a_1)^2 \rightarrow 4p$ resonances, and continuously decreases above the Cl 1s continuum.

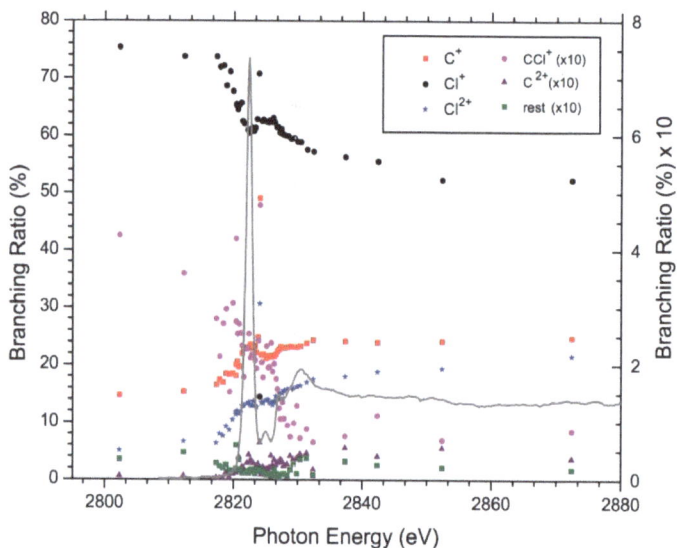

Fig. (7a). Partial ion yields for all observable fragment ions in the vicinity of the Cl $1s$ ionization threshold of CCl_4. Solid line, total ion-yield spectrum of the CCl_4 molecule, for comparison.

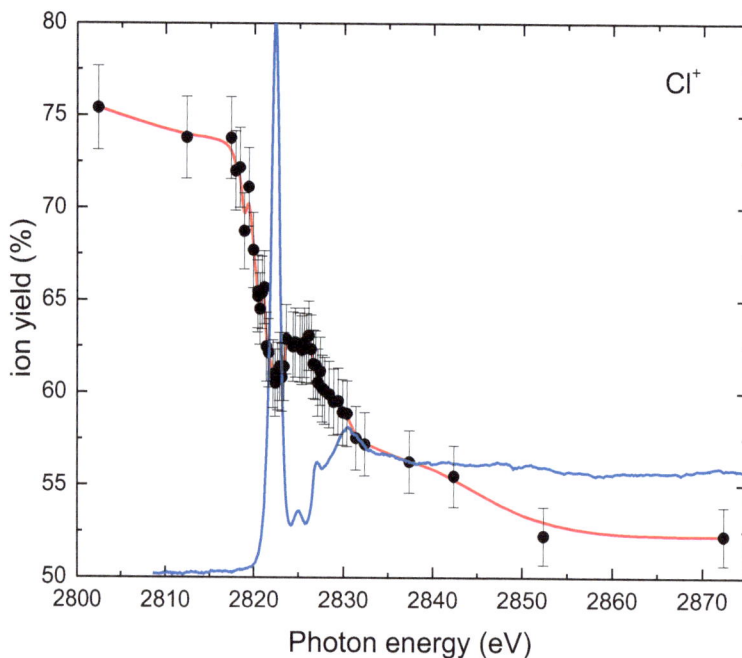

Fig. (7b). Partial ion yield of the Cl^+ fragment as a function of the photon energy around the Cl 1s edge. The photoabsorption spectrum (blue line) is also shown for the sake of comparison.

Figs. (**8** and **9**) show the partial ion yield of the C^+ and Cl^+ fragments, respectively, as a function of the photon energy around the Cl 1s edge. The photoabsorption spectrum (blue line) is also shown for the sake of comparison.

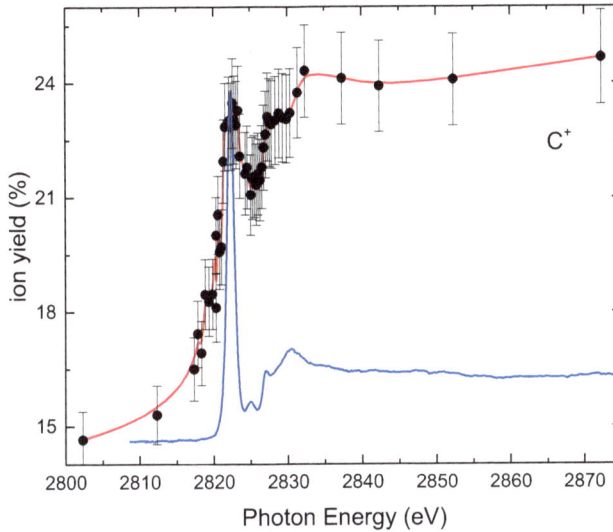

Fig. (8). Partial ion yield of the C^+ fragment as a function of the photon energy around the Cl 1s edge. The photoabsorption spectrum (blue line) is also shown for the sake of comparison.

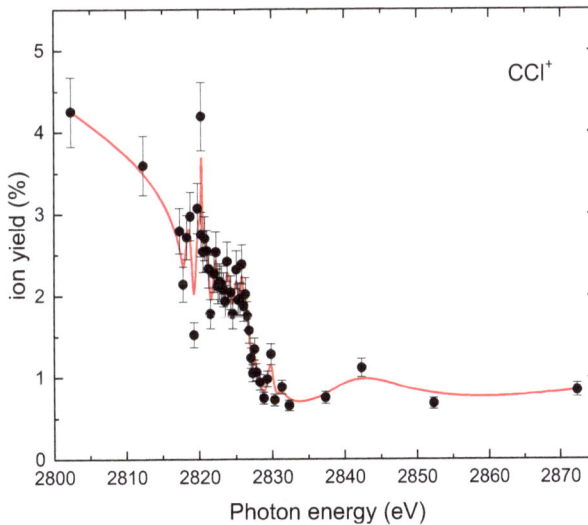

Fig. (9). Partial ion yield of the CCl^+ fragment as a function of the photon energy around the Cl 1s edge.

The C^+ branching ratio exhibits a roughly complementary behavior in relation to Cl^+ ion. Its relative intensity increases non linearly from ~ 15% above the edge to ~ 24% at Cl 1s continuum. It exhibits a maximum at the $(1t_2)^6(1a_1)^2 \rightarrow 7a_1$ resonance, probably due to the role played by the resonance Auger decay, and a minimum around the $(1t_2)^6(1a_1)^2 \rightarrow 8t_2$ resonance. The only molecular fragment observed in the present energy range is the CCl^+ ion (4% bellow the edge and 1% above it). Two doubly-charged fragments, Cl^{2+} and C^{2+}, can be observed with relevant statistics (> 0.5%). Both branching ratios increases non linearly as a function of the photon energy. Fig. (**4**) also shows the sum of the corresponding branching ratios for the rest of fragments, which do not reach 0.6%. Partial ion yield of the C^{2+} fragment as a function of the photon energy around the Cl 1s edge.

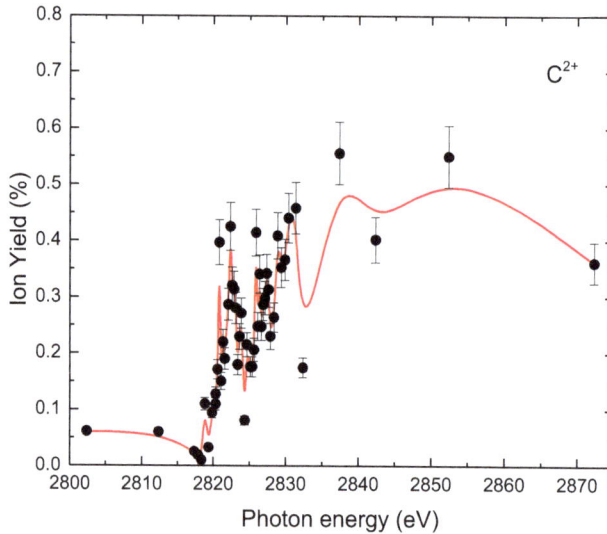

Fig. (10). Partial ion yield of the C^{2+} fragment as a function of the photon energy around the Cl 1s edge.

We now discuss an analysis of double ionization, above the Cl 1s ionization potential, based on a half-collision model [27]. Double photoionization can be treated as a sequence of two independent processes. It begins with the photoionization of a single electron. This photoelectron then causes electron-impact ionization of the remaining system. This model can be seen as a further step of the simple picture formerly proposed by Samson [28] for the double ionization of atoms: the ionization of the second electron should be similar to the electron impact ionization by the photoelectron which leaves the molecule at a low speed. Thus, the ratio of double to single ionization at low excess energy range (the photon energy minus the Cl 1s ionization potential), which is proportional to the probability for ionization of the second electron upon photoionization, should be proportional to the electron–impact single ionization cross section of the singly charged remaining ion. Therefore, by picturing electron-electron scattering, electron correlation in the double continuum can be understood. There are, nevertheless, significant differences to electron-impact ionization; as the photoelectron (the projectile) is originally localized inside the molecule near the nucleus, the electron correlation 'on the way out' resembles to a 'half-collision.' This mechanism is sometimes denoted as knock-out (KO) or two-step 1 (TS1) [5]. Additionally, while the electron-impact single ionization cross section decreases as lnv/v^2, where v is the electron speed, the double-ionization cross section should come together to the `shake-off' (SO) limit [4]. In the KO empirical model, the double ionization branching ratio as a function of the photon energy is given by [29]:

$$P_{KO}(E) = P_{KO}^{\max}\left\{\cosh\left[\beta\ln\left(\frac{E - I_p^{2+}}{\Delta E_{KO}^{\max}}\right)\right]\right\}^{-\frac{1}{\beta}} \qquad \textbf{(1)}$$

where Eq. 1 represents the analytical form of the universal shape function for electron impact ionization of H-like ions of Aichele *et. al.* [30]. Fig. (**5**) shows the Cl^{2+} branching ratio as a function of the photon energy. A set of Gaussian functions representing the resonances (see Fig. **1**) added to a constant background due to Cl 2p, Cl 2s and C 1s contributions is also shown. The KO model (Eq. 1) is shown above the Cl 1s ionization potential. Fig. (**11**) shows the Cl^{2+} branching ratio as a function of the photon energy

CONCLUSIONS

We have studied the fragmentation of the CCl_4 molecule, following core-shell photoexcitation and photoionization in the neighborhood of the Cl *1s* edge. The absorption yield has been ascribed to a new assignment of the spectral features. The structure that appears above the Cl 1s continuum, has been ascribed in terms of the existing connection with electron-CCl_4 scattering through experimental data and calculations for low-energy electron-molecule cross sections. Furthermore, the production of the doubly ionized Cl^{2+} as a function of the photon energy has been analysed in the terms of a simple and appealing physical picture, the half-collision model.

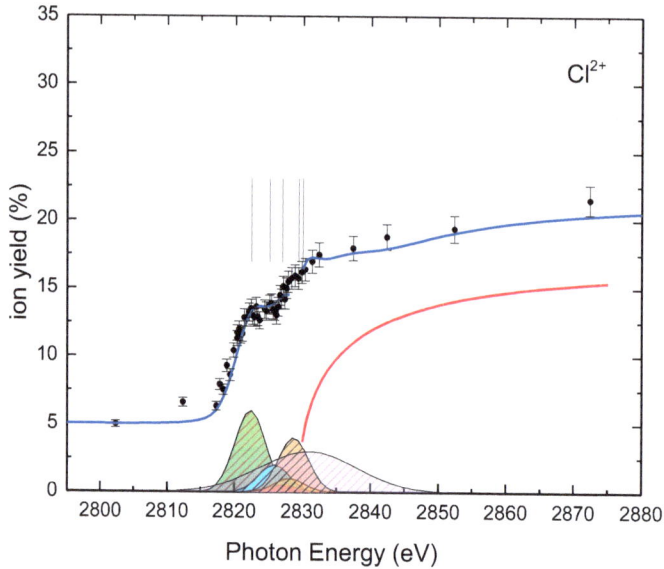

Fig. (11). Cl^{2+} branching ratio as a function of the photon energy. Red line: set of Gaussian functions representing the resonances (see Fig. **4**) added to a constant background due to Cl 2p, Cl 2s and C 1s contributions. Dotted line: KO model (Eq. 1); Blue line above Cl 1s Ip: sum of KO process, constant background due to Cl 2p, Cl2s, and C 1s contributions, and the resonances around the Cl 1s edge.

CONSENT FOR PUBLICATION

Not applicable.

CONFLICT OF INTEREST

The author (editor) declares no conflict of interest, financial or otherwise.

ACKNOWLEDGEMENTS

This work is supported in part by CNPq (Brazil).

REFERENCES

[1]   O. Travnikova, T. Marchenko, G. Goldsztejn, K. Jänkälä, N. Sisourat, S. Carniato, R. Guillemin, L. Journel, D. Céolin, R. Püttner, H. Iwayama, E. Shigemasa, M.N. Piancastelli, and M. Simon, "Hard-X-Ray-Induced Multistep Ultrafast Dissociation", *Phys. Rev. Lett.,* vol. 116, no. 21, p. 213001, 2016. [http://dx.doi.org/10.1103/PhysRevLett.116.213001] [PMID: 27284654]

[2]   T. Schneider, P.L. Chocian, and J.M. Rost, "Separation and identification of dominant mechanisms in double photoionization", *Phys. Rev. Lett.,* vol. 89, no. 7, p. 073002, 2002. [http://dx.doi.org/10.1103/PhysRevLett.89.073002] [PMID: 12190518]

[3]   T.D. Thomas, Transition from Adiabatic to Sudden Excitation of Core Electrons. *Phys. Rev. Lett.,* vol. 52, p. 419, 1984. [http://dx.doi.org/10.1103/PhysRevLett.52.417]

[4]   A.C.F. Santos, and D.P. Almeida, On the shake-off probability for atomic systems. *J. Electron Spectrosc. Relat. Phenom.,* vol. 210, pp. 1-4, 2016.
[http://dx.doi.org/10.1016/j.elspec.2016.04.005]

[5]   R.D. DuBois, A.C.F. Santos, and S.T. Manson, Empirical formulas for direct double ionization by bare ions: Z = −1 to 92. *Phys. Rev. A,* vol. 90, p. 052721, 2014.
[http://dx.doi.org/10.1103/PhysRevA.90.052721]

[6]   R. Wehlitz, Double photoionization of hydrocarbons and aromatic molecules *J. Phys. At. Mol. Opt. Phys.,* vol. 35, p. 61, 2002.

[7]   M.N. Piancastelli, The neverending story of shape resonances. *J. Electr. Spectr. Rel. Phen.,* vol. 100, p. 167, 1999.
[http://dx.doi.org/10.1016/S0368-2048(99)00046-8]

[8]   J.L. Dehmer, and D. Dill, Shape-resonance-enhanced nuclear motion effects in electron-molecule scattering and molecular photoionization, Argonne National Laboratory (ANL), Electron-Molecule Collisions Satellite Meeting, XI ICPEAC, Argonne, IL (United States), CONF-790861-6 (1979).

[9]   G.C. Schwartz, and P.M. Schaible, Rate equation for desorbing particles. *J. Vac. Sci. Technol.,* vol. 16, p. 54, 1979.
[http://dx.doi.org/10.1116/1.569867]

[10]   K.T. Lu, J.M. Chen, J.M. Lee, C.K. Chen, T.L. Chou, and H.C. Chen, State-specific dissociation enhancement of ionic and excited neutral photofragments of gaseous CCl^4 and solid-state analogs following Cl 2p core-level excitation. *New J. Phys.,* vol. 10, p. 053009, 2008.
[http://dx.doi.org/10.1088/1367-2630/10/5/053009]

[11]   M. de Simone, M. Coreno, M. Alagia, R. Richter, and K.C. Prince, Inner shell excitation spectroscopy of the tetrahedral molecules CX_4 (X = H, F, Cl). *J. Phys. At. Mol. Opt. Phys.,* vol. 35, p. 61, 2002.
[http://dx.doi.org/10.1088/0953-4075/35/1/305]

[12]   A.C.F. Santos, J.B. Maciel, and G.G.B. De Souza, Valence and shallow core fragmentation of CCl_4 molecule. *J. Electr. Spectr. Rel. Phen.,* vol. 156, p. LIX, 2007.

[13]   B.E. Cole, R.N. Dexter, and J. Quant, Photoabsorption cross sections for chlorinated methanes and ethanes between 46 and 100 Å. "Spectry", *Radiative Transfer.,* vol. 19, p. 303, 1978.
[http://dx.doi.org/10.1016/0022-4073(78)90063-8]

[14]   "G. O'Sullivan", Chlorine L-edge absorption in CCl_4 and CCl_2F_2. *J. Phys. At. Mol. Opt. Phys.,* vol. 15, p. 2385, 1982.

[15]   W. Zhang, T. Ibuki, and C.E. Brion, Absolute dipole differential oscillator strengths for inner shell spectra from high resolution electron energy loss studies of the freon molecules CF_4, CF_3Cl, CF_2Cl_2, $CFCl_3$ and CCl_4. *Chem. Phys.,* vol. 160, p. 435, 1992.
[http://dx.doi.org/10.1016/0301-0104(92)80011-J]

[16]   R.C.C. Perera, P.L. Cowan, D.W. Lindle, R.E. LaVilla, T. Jach, and R.D. Deslattes, "Molecular-orbital studies via satellite-free x-ray fluorescence: Cl K absorption and K-valence-level emission spectra of chlorofluoromethanes", *Phys. Rev. A,* vol. 43, no. 7, pp. 3609-3618, 1991.
[http://dx.doi.org/10.1103/PhysRevA.43.3609] [PMID: 9905448]

[17]   M. Simon, L. Jounel, R. Guillemin, and W.C. Stolte, "I, Minkiv, F. Gel'mukhamov, P. Salek, H. Agren, S. Carniato, R, Taieb, A. C. Hudson, and D. W. Lindle. Elastic peak of K shell excited HCl molecule: Comparison HCl–DCl—Experiment and theory", *J. Electron Spectrosc. Relat. Phenom.,* vol. 155, p. 91, 2007.
[http://dx.doi.org/10.1016/j.elspec.2006.11.007]

[18]   W.C. Wiley, and I.H. McLaren, Time□of□Flight Mass Spectrometer with Improved Resolution *Rev. Sci. Instrum.,* vol. 26, p. 1150, 1950.
[http://dx.doi.org/10.1063/1.1715212]

[19]   J.A.R. Samson, W.C. Stolte, Z-X. He, J.N. Cutler, Y. Lu, and R.J. Bartlett, Double photoionization of helium. *Phys. Rev. A,* vol. 57, p. 1906, 1998.
[http://dx.doi.org/10.1103/PhysRevA.57.1906]

[20]   K.T. Lu, J.M. Chen, J.M. Lee, S.C. Haw, T.L. Chou, S.A. Chen, and T.H. Chen, Core-level anionic photofragmentation of gaseous CCl_4 and solid-state analogs. *Phys. Rev. A,* vol. 80, p. 033406, 2009.
[http://dx.doi.org/10.1103/PhysRevA.80.033406]

[21]   M. Braun, S. Marienfeld, M-W. Ruf, and H. Hotop, High-resolution electron attachment to the molecules CCl_4 and SF_6 over extended energy ranges with the (EX)LPA method *J. Phys. At. Mol. Opt. Phys.,* vol. 42, p. 125202, 2009.
[http://dx.doi.org/10.1088/0953-4075/42/12/125202]

[22]   R.K. Jones, Absolute total cross sections for the scattering of low energy electrons by CCl_4, CCl_3F, CCl_2F_2, $CClF_3$, and CF_4. *J. Chem. Phys.,* vol. 84, p. 813, 1986.
[http://dx.doi.org/10.1063/1.450580]

[23]   P.D. Burrow, A. Modelli, N.S. Chiu, and K.D. Jordan, Temporary negative ions in the chloromethanes $CHCl_2F$ and CCl_2F_2: Characterization of the σ* orbitals. *J. Chem. Phys.,* vol. 77, p. 2699, 1982.
[http://dx.doi.org/10.1063/1.444103]

[24]   R. Curik, F.A. Gianturco, and N. Sann, Electron and positron scattering from halogenated methanes: a comparison of elastic cross sections. *J. Phys. At. Mol. Opt. Phys.,* vol. 33, p. 615, 2000.
[http://dx.doi.org/10.1088/0953-4075/33/4/302]

[25]   A.P.P. Natalense, M.H.F. Bettega, L.G. Ferreira, and M.A.P. Lima, "Low-energy electron scattering by CF4, CCl4, SiCl4, SiBr4, and SiI4", *Phys. Rev. A,* vol. 52, no. 1, pp. R1-R4, 1995.
[http://dx.doi.org/10.1103/PhysRevA.52.R1] [PMID: 9912322]

[26]   G.M. Moreira, A.S. Barbosa, D.F. Pastega, and M.H.F. Bettega, Low-energy electron scattering by carbon tetrachloride. *J. Phys. At. Mol. Opt. Phys.,* vol. 49, p. 035202, 2006.
[http://dx.doi.org/10.1088/0953-4075/49/3/035202]

[27]   W.C. Stolte, V. Jonauskas, D.W. Lindle, M.M. Sant'Anna, and D.W. Savin, inner-shell photoionization studies of neutral atomic nitrogen. *Astrophys. J.,* vol. 818, p. 149, 2016.
[http://dx.doi.org/10.3847/0004-637X/818/2/149]

[28]   J.A.R. Samson, "Proportionality of electron-impact ionization to double photoionization", *Phys. Rev. Lett.,* vol. 65, no. 23, pp. 2861-2864, 1990.
[http://dx.doi.org/10.1103/PhysRevLett.65.2861] [PMID: 10042717]

[29]   J. Hoszowska, J-Cl. Dousse, W. Cao, K. Fennane, Y. Kayser, and M. Szlachetko, Double K-shell photoionization and hypersatellite x-ray transitions of $12 \leq Z \leq 23$ atoms. *Phys. Rev. A,* vol. 82, p. 063408, 2010.
[http://dx.doi.org/10.1103/PhysRevA.82.063408]

[30]   K. Aichele, U. Hartenfeller, D. Hathiramani, G. Hofmann, V. Schäfer, M. Steidl, M. Stenke, E. Salzborn, T. Pattard, and J.M. Rost, Electron impact ionization of the hydrogen-like ions B^{4+}, C^{5+}, N^{6+} and O^{7+}. *J. Phys. B,* vol. 31, p. 2369, 1998.
[http://dx.doi.org/10.1088/0953-4075/31/10/023]

CHAPTER 7

Multiple Scattering in Electron Rutherford Scattering Spectroscopy

K. Tőkési* and **D. Varga**

Institute for Nuclear Research, Hungarian Academy of Sciences (MTA Atomki), H–4001 Debrecen, P.O. Box 51, Hungary, EU

Abstract: We present a theoretical description of the spectra of electrons elastically scattered from various samples. The analysis is based on very large scale Monte Carlo simulations of the recoil and Doppler effects in reflection and transmission geometries. Besides the experimentally measurable energy distributions the simulations give many partial distributions separately, depending on the number of elastic scatterings (single, and multiple scatterings of different types). Furthermore, we present detailed analytical calculations for the main parameters of the single scattering, taking into account both the ideal scattering geometry, *i.e.* infinitesimally small angular range, and the effects of the real, finite angular range used in the measurements. The effect of the multiple scattering on intensity ratios, peak shifts and broadening, are shown. We show results for multicomponent and double layer samples. Our Monte Carlo simulations are compared with experimental data. We found that our results are in good agreement with the experimental observations.

Keywords: Elastic scattering, Monte Carlo simulation, Multiple scattering, Recoil and Doppler effects, Reflection and transmission geometries.

INTRODUCTION

In recent times the recoil energies of scattered electrons for atoms with large mass differences can be well resolved by using an energetic electron beam in the range of a few keV [1 - 6] to a few tens of keV, and with large scattering angles in the measurements [7 - 18]. This technique is called as Electron Rutherford Backscattering Spectroscopy (ERBS), which relies on the quasi-elastic electron-atom scattering. In this case, we take advantage of the fact that the energy of the elastically scattered electrons is shifted from the primary values, due to the momentum transfer between the primary electron and the target atoms (recoil

---

* **Corresponding author K. Tőkési:** Institute for Nuclear Research, Hungarian Academy of Sciences (MTA Atomki), H–4001 Debrecen, P.O. Box 51, Hungary, EU; Tel: +36 52 509-245; E-mail: tokesi@atomki.mta.hu

Antônio Carlos Fontes dos Santos (Ed.)

effect), and thereby the peak, due to electrons scattered elastically, splits into component peaks, which can be associated with the electrons scattered mainly from different target atoms of the sample, respectively. Furthermore, the thermal motion of the scattering atoms causes broadening in the primary electron energy distribution, usually referred to as Doppler broadening. So, from the accurate determination of the full width at half maximum (FWHM) of the peaks, the average kinetic energy of the atoms in a solid can be determined. Moreover, from the accurate peak shape analysis we can determine the Compton profile [8, 19] or we can prognosticate different fine interaction processes, such as, final state interactions.

Observation of the hydrogen peak is either a challenging or impossible task for the conventional electron spectroscopy. Hydrogen was observed earlier in electron scattering experiments at high energy using transmission geometry and formvar film [8]. This technique was used for the determination of the hydrogen content at surfaces [2 - 4, 20]. These investigations help us to improve our understanding of the processes that involve the presence of H atoms at surfaces in polymers, carbon based hard coatings or new H storage materials. These later studies can be closely related to discovering new energy cells for new generation cars operating with hydrogen and producing simple water instead of toxic environment depleting gases. This procedure has also been used to check surface degradation in several polymers [5]. Filippi and Calliari extended this strategy for the quantification of H at more complex surfaces, containing other atoms like O [6].

In the first part of this chapter we present detailed theoretical as well as experimental studies for the detection of hydrogen by analyzing the spectra of electrons backscattered elastically from polyethylene ($(CH_2)_n$). Monte Carlo calculations were carried out in order to simulate the spectra of backscattered electrons having primary energies between 1 and 5 keV. The number of primary electrons was extremely large, 5×10^{10}, because of our desire to reach a very small statistical error for comparison with the experimental data. We note, that many X-ray photoelectron spectrometers also have an electron gun. And using this technique the hydrogen content of various samples can be determined. This can not be seen in the photoelectron spectra.

Following the same principles but using higher electron kinetic energies extend the information depth during the investigations. Besides the fundamental research interest, applying electrons with relativistic energies *(E > 10 keV)*, can highlight important technical applications. For example, from the measurement of the relative peak intensities we can estimate the thickness of the layer from where the signal originates. So far, many experiments have been done also in the relativistic electron energy range. Vos and Went [12, 14] proposed to study surface atomic composition and vibrational properties. They also used various kinds of specimens, such as bulk elements [10, 11], overlayer/substrate systems [10 - 13], alloy and composite bulks [11, 14], to measure the recoil energies, the Doppler broadening of elastic peaks, and especially the peak intensity ratios between different atoms.

In the second half of this chapter accurate Monte Carlo simulations are presented for double layer samples in order to simulate the spectrum of elastically scattered electrons having 40 keV primary energy. The analysis was based on the Monte Carlo simulations of the recoil and Doppler effects in reflection and transmission geometries of the scattering at a fixed scattering angle of 44.3°. The relativistic correction was also taken into account. Besides the experimentally measurable energy distributions we present many partial distributions separately, depending on the number of elastic scatterings (single, and multiple scatterings of different types). Furthermore, we present detailed analytical calculations for the main parameters of the single scattering taking into account both the ideal scattering geometry, *i.e.* infitesimally small angular range, and the effect of the real, finite angular range used in the measurements. We show our results for intensity ratios, peak shifts and broadenings in four cases of measurement geometries and layer thicknesses. The effects of the multiple and mixed scatterings on the parameters of the elastic peak are also investigated. Finally, we make an attempt to compare our Monte Carlo results with the experimental results measured by Vos and Went [16]. Here we would like to note that some of the input data of our Monte Carlo calculation may differ slightly from the data realized in the given experiments (see, for example, the real solid angle during the measurement or the accurate knowledge of the thickness of the layers).

THEORY

Energy Shift and Broadening Due to Elastic Scattering

Within the classical model for single elastic electron scattering on free atoms, the energy transferred to an atom of zero kinetic energy is [7]:

$$E_{r0} = \frac{4m}{M} E_0 \sin^2(\theta_0 / 2), \tag{1}$$

where θ_0 is the angle of scattering, m and E_0 are the mass of the electron and its energy before scattering and M is the mass of the scattering atom. In the case of a scattering atom having a kinetic energy ε, the recoil energy is from energy and momentum conservation [1]:

$$E_r = \frac{2m}{M} E_0 \left[1 - \cos\theta_0 + \sqrt{\frac{M\varepsilon}{mE_0}} \left(\cos\vartheta - \cos\theta_0 \cos\vartheta - \sin\theta_0 \sin\vartheta \cos\varphi \right) \right], \quad (2)$$

where the angles ϑ and φ are characterizing the direction of the velocity of the scattering atom, relative to the direction of the velocity of the primary electron and to the plane of the scattering, respectively. Assuming a fixed value of θ_0, an isotropic distribution for the velocity directions of the atoms and the Maxwell-Boltzmann thermal distribution for their energy, a Gaussian energy distribution is obtained for the elastic peak, with its maximum at E_{r0} and full width at half maximum (ΔE_r) [1]:

$$\Delta E_r = 4\sqrt{\frac{2}{3}} \, E_{r0} \, \bar{\varepsilon} \ln 2 \qquad (3)$$

where $\bar{\varepsilon}$ is the mean value of kinetic energy. It should be emphasized here, that opposite to the statement of Kwei *et al.* [21], the Gaussian shape of the elastic peak with E_{r0} energy position of its maximum is not an assumption, but the consequence of the Maxwell-Boltzmann thermal distribution assumed for the energy of scattering atoms. From Eqs. (1) and (3) it can be seen, that in the case of the elastic peak in the spectrum of electrons backscattered from a solid surface, the energy shift of the elastic peak is proportional to E_0/M, while its ΔE_r (Doppler) broadening to $\sqrt{E_0/M}$. It can also be proved easily that in the case of large angle electron backscattering the high energy resolution measurement of the elastic peak has special advantages not only for achieving a complete separation of the elastic and inelastic parts in the spectra (important for determining the inelastic mean free path of the electrons in solids, see [22] and references therein), the energy position and shape of the elastic peak contains information on the chemical composition and/or the cleanliness of the surface layers of the sample as well [23].

In the case of a clean, homogeneous sample consisting of a single component and a given E_0 primary electron energy, the experimentally derived energy width of the elastic peak includes the material and temperature dependent Doppler broadening, the instrumental contributions due to the finite energy resolution of the spectrometer and the energy distribution of the primary beam. In addition, however, the measured energy width of the elastic peak can include contributions electrons that have suffered very small energy losses due to inelastic processes (as a consequence of *e.g.* phonon excitation or other excitation processes) which are unseparable from the elastic peak [24]. The Doppler broadening is also dependent on the $\Delta\theta_0$ angular range determined by the spectrometer and the electron gun, as well as by the number of elastic scattering events when multiple scattering takes place. The formulas above are concerned with single scattering, however, since the contribution of multiply scattered electrons can be very large in the primary beam energy region of a few keV studied by us for investigating this effect, we have performed Monte Carlo model calculations to derive the increase of the elastic peak broadening as a consequence of multiple scattering for homogeneous sample with two components.

Monte Carlo Simulation

Monte Carlo simulation of electron transport in solids is based on the stochastic description of scattering processes. Electron penetration is approximated by a classical zigzag trajectory. Details of the calculations are given elsewhere [25]. The scattering point is where the electron changes its direction and/or energy. In our calculations both the elastic and inelastic scattering events were taken into account. For the case of the first inelastic collision the calculations were stopped. Particular values of scattering angles of electrons in an individual event are realized by random numbers following the angular differential elastic cross sections of the given target atom. After each elastic scattering the recoil energy was calculated according to Eq. 2. Fig. (**1**) shows the geometric configuration used in the calculations.

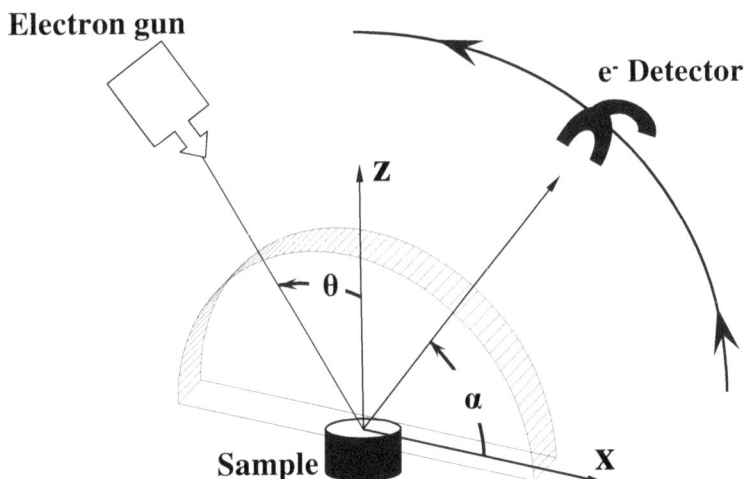

Fig. (1). Schematic view of the geometric configuration of the calculation.

By the help of our recently developed Monte Carlo code we are able to calculate the angular distributions of backscattered electrons in the *xz*-plane (see Fig. **1**) as a function of the angle of emission α at the given initial angle of incident beam (θ).

Elastic Scattering

The partial expansion method was used to describe the differential and total cross sections for elastic scattering. The elastic scattering of relativistic particles is described by the direct $f(\theta)$ and spin-flip $g(\theta)$ scattering amplitudes. The relativistic differential cross section per unit solid angle is given by

$$\frac{d\sigma_e}{d\theta} = |f(\theta)|^2 + |g(\theta)|^2 ,\tag{4}$$

where θ denotes the scattering angle. Details of the calculations can be found in ref [26]. The scattering angle θ is calculated using the random number $R_1\epsilon\,(0, 1)$ satisfying the relation:

$$R_1 = \frac{2\pi}{\sigma_e} \int_0^\theta \frac{d\sigma_e(E,\theta)}{d\Omega} \sin\theta\, d\theta ,\tag{5}$$

where θ_e is the total elastic scattering cross-section. A further random number $R_2\,\square\,(0, 1)$ selects the azimuthal angle

$$\phi = 2\pi R_2\tag{6}$$

after the elastic collision. The mean free path for elastic scattering λ_e is given by

$$\lambda_e = \frac{A}{N_a\rho\sigma_e} ,\tag{7}$$

where A is the atomic weight of the target material, ρ is the density, N_a is the Avogadro's number.

Inelastic Scattering

The mean free path for inelastic scattering of electrons within a solid path was described with the predictive formula of Tanuma *et al.* [27]:

$$\lambda_{in} = \frac{E}{E_p^2 \{\beta \ln(\gamma E) - C/E + D/E^2\}}, \tag{8}$$

where E is the electron energy in eV, E_p is the free electron plasmon energy in eV, β, γ, C, and D are fitting parameters.

QUANTITATIVE ANALYSIS OF ELECTRONSPECTRA FROM MULTICOMPONENT SAMPLE – THE CASE OF POLYETHYLENE SAMPLE

Characterization of the Model

If the thickness of the sample significantly exceeds the inelastic mean free path, the 30-40% of the intensity of the quasi-elastic peak arises from the multiple scattering in reflection mode [28]. In our recent studies the sample can be treated as a "semi-infinite sample". In extreme cases the contribution from the multiple scattering should even exceed 80%. For the case of multicomponent samples, the estimation of the multiple scattering effects is a very hard task. This fact is especially true when we are also interested in the energy distribution of the backscattered electrons. Namely, if the sample contains atoms with different masses, one part of the multiple scattering is come from the mixed collisions, *i.e.* the electron scattered on atoms with different masses and the resultant energy loss distribution can not be described with the single Gauss distribution. Therefore, for mimicking the prospective electron energy distributions backscattered from a polyethylene (PE) sample we used Monte Carlo simulations. With the help of our Monte Carlo code we are able to calculate the angular distributions of backscattered electrons in the xz-plane (see Fig. **1**) as a function of the angle of emission α at the given initial angle of incident beam (θ). In our recent experiment $\alpha = 115°$ and $\theta = 75°$, and accordingly $\theta_0 = 130°$. The applied model used the simple impulse approximation. The main aspects of the model are as followings:

- The sample is semi-infinite, homogeneous and amorphous.
- The electron motion in the sample is treated as zigzag trajectory. If the electron collides with a moving atom, after the elastic collision its energy exchange, according to Eq. 2. can be calculated by the following formula:

$$E_{rl} = \frac{2m}{M_l} E_0 \left[1 - \cos\theta_0 + \sqrt{\frac{M_l \varepsilon_l}{m\varepsilon_0}} \left(\cos\vartheta_l - \cos\theta_0 \cos\vartheta_l - \sin\theta_0 \sin\vartheta_l \cos\varphi_l \right) \right], \tag{9}$$

where m, E_0 and θ_0 are the mass, the energy and the scattering angle of the electron respectively. l characterizes the atom where the elastic scattering happened. In our present case l is either C or H. M_l, ε_l, are the mass and energy of the given atom, and the angles ϑ_l and φ_l are characterizing the direction of the velocity of the scattering atom, relative to the direction of the velocity of the primary electron and to the plane of the scattering, respectively.

- The partial expansion method was used to describe the differential and total cross sections for elastic scattering [26]. In our calculations atomic target and Dirac-Hartree-Fock- Slater wave functions were used. This is the more general combination in similar calculations and energy range like EPES method obtaining the inelastic mean free path [29, 30].

The mean free path for inelastic scattering of electrons within a solid was described with the predictive formula of Tanuma *et al.* (see Eq. 8). For polyethylene the parameters are the followings [27]: $\beta = 0.0188$ eV^{-1}Å^{-1}, $\gamma = 0.169$ eV^{-1}, $C = 1.33$ Å^{-1} and $D = 20.7$ eVÅ^{-1}. The calculation of the inelastic mean free path is based on the experimental optical data.

For a given scattering event, random numbers defined the nature of the collision (elastic or inelastic), the type of the atom where the electron scatters (C or H), the scattering angles, the kinetic energy of the selected target (recoiled) atom, and the instantaneous direction of the selected target atom (ϑ_1, φ_1). The selection of the collision type is based on the atomic concentration and cross sections. The instantaneous kinetic energy of the target atom is selected from the Maxwell-Boltzmann type distributions, where the average kinetic energies ($\bar{\varepsilon}$.) are described independently for the two components.

- The velocity distributions of the atoms are isotropic.
- For the case of the investigated angular range, (in our preset studies this is $\theta_0 = 130° \pm 5°$) the energy loss of the elastically scattered electrons determined and the energy loss spectra is stored.
- Beside the partial loss functions suffered 1, 2...50 elastic collisions, many other distributions can be analyzed based on the stored data of the Monte Carlo code. For example, the partial loss functions are saved as a function of the number of elastic scatterings on C or H. We note, however, that for the study of partial distributions with suitable energy resolution and accuracy even for the peak of H, the evaluation of a huge number of primary trajectories was required.

Verification of the Calculations

The yields and the energy distributions for the single scattering can be used for testing the accuracy of our simulation procedure by comparing the results of our

Monte Carlo simulations with simply analytical formulas. The absolute yields, P_l, of the electrons scattered elastically only once either on a carbon or hydrogen atom relative to the primary electron number can be written in the following formula:

$$P_l = \frac{\cos(\theta_{out})}{\cos(\theta_{out})+\cos(\theta_{in})}\frac{\lambda_i}{\lambda_e+\lambda_i}\frac{\Delta\Omega}{\sigma_e(C)+2\sigma_e(H)}X_l\left(\frac{d\sigma_e}{d\Omega}\bigg|_{\theta=\theta_0}\right)_l, \qquad (10)$$

where λ_e is the elastic mean free path, λ_i is the inelastic mean free path, $\sigma_e(C)$ and $\sigma_e(H)$ are the total elastic cross sections of electron colliding with the carbon or hydrogen atom, $\Delta\Omega$ is the solid angle of detected electrons, $\left(\dfrac{d\sigma_e}{d\Omega}\bigg|_{\theta=\theta_0}\right)$ is the angular differential elastic cross sections at the scattering angle θ_0 for $l = C$ and H, X_l is the stoichiometry number for C and H in PE. In our simulation $X_C = 1$ and $X_H = 2$.

Table **1** shows the single elastic scattering yields obtained with Eq. (10) and with our present Monte Carlo simulations. For the case of Monte Carlo simulations, the statistical errors are shown in the brackets. Due to the large number of primary trajectories (5×10^{10}) even for the smaller intensity scatterings on H, the statistical errors are below 1.4%. The results of our Monte Carlo simulations are in good agreement with the results provided by Eq. (10). The average deviation is 0.3% and 0.5% for carbon and for hydrogen, respectively. Therefore, the scattering ratio is also predicted reasonably well. On average this is 0.5% for the present primary energies. From the Eq. (10) we can write:

$$\frac{P_{1H}}{P_{1C}} = 2\frac{d\sigma}{d\Omega}(H)\bigg/\frac{d\sigma}{d\Omega}(C). \qquad (11)$$

According to the Table **1**, the angular differential cross section ratio between H and C varied between 4.9% and 5.2%. This is about 10% smaller than the expected value of 2/36 (5.56%) based on the well know Z^2 law. Using our Monte Carlo code, various electron energy loss distributions of elastic peaks can be obtained. We note that the calculations were stopped at the first inelastic collision. Let us denote these partial distributions by C^iH^j, where the index shows the number of elastic collisions on the given types of atoms. The shape of two distributions (C^1H^0 and C^0H^1) among many can be verified. From the property of our model, the Gaussian function must describe the shape of the single scattering distributions due to the Doppler effect. The shift of these distributions compared to the primary energy E_0 can be written as:

$$E_{0rl} = \frac{4m}{M_l} E_0 \sin^2(\theta_0 / 2),$$ (12)

and the full width at half maximum (FWHM) of the distributions can be characterized with the following formula:

$$\Delta E_{rl} = \sqrt{8 \ln 2} \sqrt{\frac{4}{3} E_{0rl} \bar{\varepsilon}_l}.$$ (13)

Table 1. Comparison of single elastic scattering yields obtained with Eq. (10) and with our present Monte Carlo simulations.

E_0	$P_{1C}(10^{-6})$		$P_{1H}(10^{-7})$		$P_{1H}/P_{1C}(10^{-2})$	
(keV)	Equation (10)	Monte Carlo ±(%)	Equation (10)	Monte Carlo ±(%)	Equation (10)	Monte Carlo ±(%)
1	15.1024	15.1566(0.11)	6.8598	6.848(0.5)	4.542	4.518(0.7)
2	6.2213	6.2297(0.18)	3.0420	3.022(0.8)	4.890	4.851(1.0)
3	3.7847	3.7744(0.23)	0.1920	1.897(1.0)	5.052	5.026(1.2)
4	2.6849	2.6958(0.27)	1.3817	1.390(1.2)	5.146	5.156(1.5)
5	2.0653	2.0728(0.31)	1.0756	1.076(1.4)	5.208	5.192(1.7)

The area of the Gaussian peaks, normalized to the total number of primary electron numbers entering the PE sample, is given by P_1 and described by Eq. (10). The C^1H^0 and C^0H^1 energy loss distributions were calculated and fitted by Gaussian peaks. The calculated energy shifts, FWHMs and peak areas for the single scatterings are in very good agreement with the analytical expressions. The deviations from the analytical values are 0.3%, 0.8%, 0.2% for the lower intensity C^0H^1 cases, respectively. For the case of C^1H^0 these deviations are even smaller. We note also that in verifying this observation the solid angle $\Delta\Omega$ was varied. We calculated the energy loss distributions when $\Delta\Omega$ was ±2°, ±3°, ±5°. We found that the results, within the margin of error, were the same. Therefore, in the following we will show our Monte Carlo results for $\Delta\Omega = \pm 5°$.

Results of the Monte Carlo Simulations

Fig. (**2**) shows the 3D color plot of the calculated electron intensity backscattered elastically from polyethylene. The separation between the carbon and hydrogen peak is clearly seen as a function of emission angle.

Table **2** shows the electron yields backscattered elastically at $\theta = 130°$ and with $\Delta\Omega = \pm5°$ solid angle as a function of the primary energy E_0. The data are given in % normalized to the total number of the backscattered electron number.

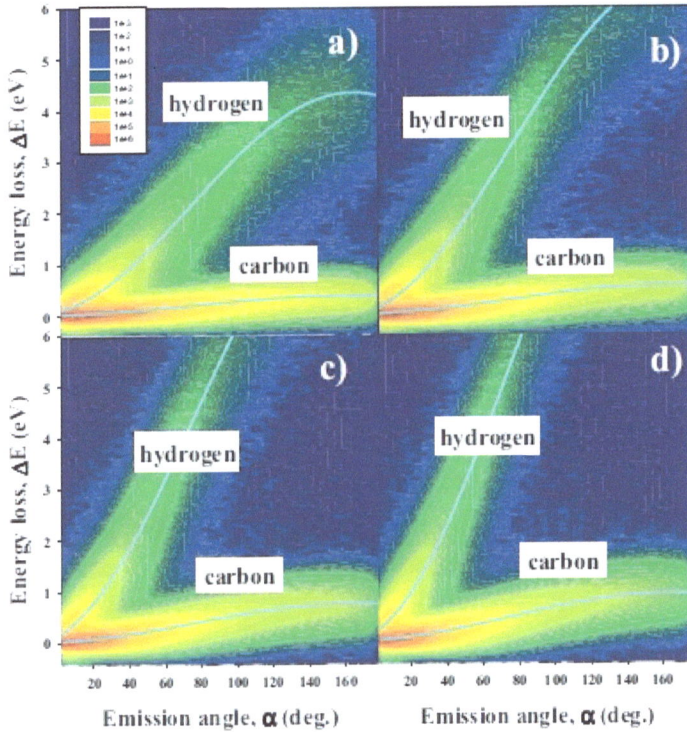

Fig. (2). Energy loss distributions of electrons backscattered elastically from polyethylene at (a) 1 keV; (b) 3 keV; (c) 4 keV; and (d) 5 keV primary energies.

The columns 2 and the 3 in Table **2** show the single scattering contributions either on carbon or hydrogen atoms. The sum of these two contributions is about 60% for each energy. Therefore, the contribution by the multiple scatterings is about 40%. The second half of Table **2** shows other yields: columns 4, 5 and 6 display the partial elastically backscattered yields when the electrons scatter only on carbon atoms, only on hydrogen atoms, and when the electron scatters both on carbon and hydrogen, respectively. The later yields have important roles in the energy loss spectra. The difference between the columns 2 and 4 shows that contributions of the multiple scattering are significant for carbon atoms. At the same time, the difference between the columns 3 and 5 are very small. This fact indicates that the yield of the multiple scattering is negligible for the hydrogen atoms.

In both cases when the electron scattered on only one type of atoms ("mono-atomic scattering" C^iH^0 and C^0H^j), the electron energy loss spectra can be fitted with Gaussian peaks reasonably well. The calculated average E_{0rl} energy shift is 0.4% smaller than that calculated with Eq. (12). At the same time, the FWHMs of the peaks were bigger by about 1-2% than it predicted by Eq. (13).

Table 2. Electron yields backscattered elastically form different type of target atoms at $\theta = 130°$ and $\Delta\Omega = \pm 5°$ solid angle.

E_0 (keV)	$C^1H^0(\%)$	$C^0H^1\ (\%)$	$C^iH^0\ (\%)$ $i \geq 1$	$C^0H^j\ (\%)$ $j \geq 1$	$C^iH^j\ (\%)$ $i \geq 1, j \geq 1$
1	54.59	2.47	88.94	2.58	8.48
2	56.08	2.72	89.50	2.84	7.66
3	57.18	2.87	89.75	2.98	7.27
4	58.27	3.00	89.97	3.11	6.92
5	58.78	3.05	90.16	3.18	6.66

These results are in agreement with our previous observations, where the energy shift for the case of single component Si and C sample was ~1.5% smaller, and the ΔE_r width was ~2.5% greater than the corresponding values calculated by the single scattering formulas [1]. The contribution of the multiple scatterings was greater for both samples than for PE.

We note, that the ratio between the yields of H and C peaks (C^0H^1/C^1H^0) calculated by the mono-atomic scattering is about 30% smaller than can be estimated from the differential cross sections. Therefore, likely the most interesting part of Table **2** is the last column, which verifies that the contribution of the mixed scattering events (C^iH^j, $i,j \geq 1$) is large. In each E_0 primary energies, the contribution of the mixed scattering events were at least double the mono-atomic scattering on H. The question is the following: What is the spectral distribution of these mixed scatterings? Because due to the mass difference between the C and H atoms, at the same scattering angle, the energy transfer should differ each other by an order of magnitude (see Eq. (12)). In our Monte Carlo simulation the average kinetic energy of the carbon atom was $\bar{\varepsilon}_C = 80$ meV and the average kinetic energy of the hydrogen atom was $\varepsilon_H = 120$ meV. These values were obtained from the preliminary evaluation of the experimental data [3]

As an example, Fig. (**3**) shows the energy loss distributions at 2 keV primary energy for single scatterings (C^1H^0, C^0H^1), for mono-atomic collisions (C^iH^0, C^0H^j), and for C^sH^s where all possible elastic collisions are included. We note that within the graphical resolution C^0H^j is equal with C^0H^1.

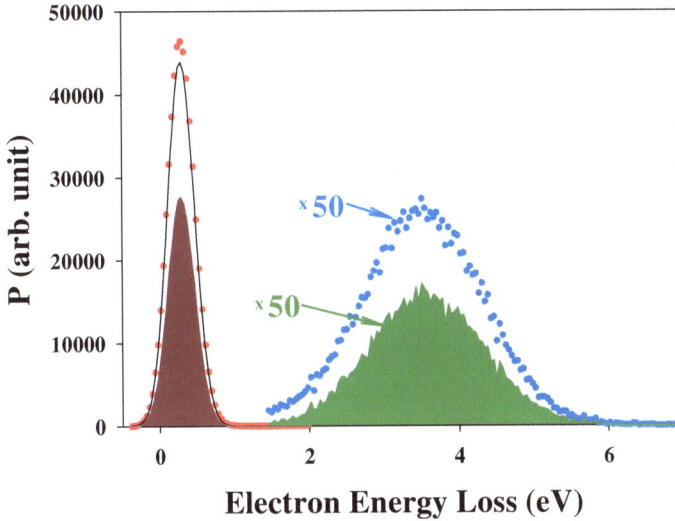

Fig. (3). Energy loss distributions at 2 keV primary energy for single scatterings (dark red shaded area: C^1H^0, green shaded area: C^0H^1), for mono-atomic collisions (solid line: C^iH^0, within the graphical resolution C^0H^1 is equal with C^0H^1), and for C^sH^s (dotted line) where all possible elastic collisions are included.

Fig. **(4)** shows the energy loss distributions of the mixed scatterings, C^iH^j, where $i, j \geq 1$ at 2 keV primary energy. While the dotted line in Fig. **(4)** displays our Monte Carlo results, the solid line shows the Gaussian curve fitting to the energy loss distributions of single scatterings (either on C or H). As for the case of single scatterings the mixed scatterings also constitute two peaks. But these peaks are a little bit wider compared to the peaks describing the single scatterings. Moreover, the peaks are asymmetric in the direction towards one another and between the peaks a continual distribution with small intensity is located. The peaks positions disagree with the energy shifts according to the mass of C and H. The deviation depends on the primary energy. The C^iH^j distributions were fitted by Gaussian peaks at 1, 2, 3, 4 and 5 keV primary energies. We found that the distance between the two peaks was smaller with $(7.7 \div 2.4)\%$, the FWHM of the C peaks was greater with $(12.9 \div 5.9)\%$, and the FWHM of the H peaks was greater with $(20.0 \div 10.0)\%$ compared with the results of the analytical formulas. Although, the deviations decrease with increasing primary energies, the ratio between the area of C and H peaks in the spectra of mixed scatterings is the same (3: 1) for different primary energies. This ratio is much more favorable for H than the single $(C^1H^0/C^0H^1 \approx 20)$ or mono-atomic scatterings $(C^iH^0/C^0H^j \approx 30)$ (see Table **2**). Therefore, the effect of the mixed scatterings in the total spectra backscattered elastically from PE (C^sH^s) is much bigger for the hydrogen peaks than for the carbon peaks. Experimentally, of course, the partial distributions cannot be seen, *only* the sum spectra for all scattering types (C^sH^s) can be investigated and compared with the theory. However, these spectra can also be fitted by two Gaussian peaks, one is related to the carbon and the other is related to the hydrogen peaks. But the results of the fitting for the peak position and the FWHM will differ from that of the results supposing only monoatomic collisions. The deviation depends on the contribution of the mixed scatterings and its spectral distributions. This causes a problem in the comparison of experimental data with the theory.

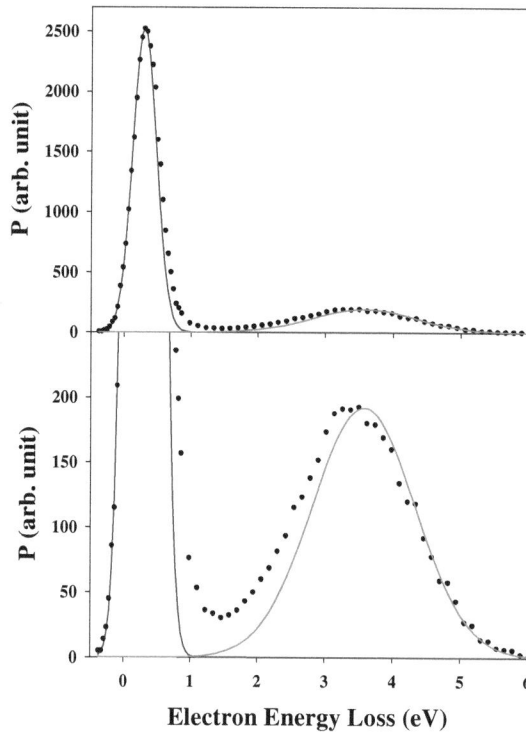

Fig. (4). Energy loss distribution of the mixed scatterings at 2 keV primary energies from PE.

Table **3** shows the parameters of the C^sH^s spectra fitted by Gaussian peaks in comparison with the results of the partial distributions when only single scatterings are taken into account. δ denotes the relative deviations in %. The first part of Table **3** shows the relative distances between the carbon and hydrogen peaks. It can be seen that the distances between the peaks become smaller due to the mixed scatterings. The difference, however, decreases with increases in primary energy. The peak widths of carbon and hydrogen peaks are compared in the second and third part of the Table **3**. While for the case of the carbon peak, the

FWHM has risen from 2% to 1.2%, for the case of the hydrogen peaks the FWHM have risen from 12.5% to 5.2%. The last part of the Table **3** displays the ratios of the peak areas between the hydrogen and carbon. In this case the δ relative deviation means the ratio between the I(H)/I(C) (from Table **3**) and the P_{1H}/P_{1C} (from Table **1**, see also Eq. (10)). Except for the case of 1 keV primary energy the relative deviations are small. The sign of the ratio also changes with increases in energy. It seems that the contribution of the mixed scatterings to the hydrogen peak compensates for the contributions of the multiple scatterings in the carbon peak. This compensation decreases the primary energy dependence of the H/C ratio. In our simulations, the compensation can be treated as a constant value, and for the case of our measurement points the average value is $(5.02\pm0.03)\times10^{-2}$. According to the differential cross sections these compensations varied from 4.89×10^{-2} to 5.21×10^{-2} in the primary energy range between 2 and 5 keV. We note, that the obtained results are strongly material and experimental dependent, so the effect of the mixed scatterings can be different for different materials. Therefore, it can cause different errors in the comparison between experiments and theory, which is difficult to interpret without detailed similar MC calculations.

Table 3. Parameters of the C^sH^s spectra fitted by Gaussian peak in comparison with the results of single scatterings.

E_0 (keV)	$E_{r0}(H) - E_{r0}(C)$ (eV)			$\Delta E_r(C)$ (eV)			$\Delta E_r(H)$ (eV)			$I(H)/I(C)(10^{-2}$		
	Eq.(12)	C^sH^s	δ (%)	Eq.(13)	C^sH^s	δ (%)	Eq.(13)	C^sH^s	δ (%)	Eq.(10)	C^sH^s	δ (%)
1	1.638	1.519	-7.3	0.298	0.304	+2.0	1.260	1.417	+12.5	4.542	4.89	+7.6
2	3.276	3.184	-2.8	0.421	0.429	+1.9	1.781	1.944	+9.2	4.890	4.96	+1.4
3	4.915	4.820	-1.9	0.516	0.523	+1.4	2.182	2.354	+7.9	5.052	5.04	-0.3
4	6.553	6.478	-1.1	0.596	0.604	+1.3	2.519	2.686	+6.6	5.146	5.05	-1.8
5	8.191	8.126	-0.8	0.666	0.674	+1.2	2.817	2.964	+5.2	5.208	5.03	-3.4

Comparison with Experiment

Polyethylene sample (Goodfellow ET311251) was used at room temperature in our experiments. The cleanliness of the sample surface was monitored by XPS analysis. As a reference specimen Cu polycrystalline metal was used. The Cu surface was cleaned by Ar^+ ion sputtering with an ion flux of 120 $\mu A\times min/cm^2$ and 2 keV kinetic energy. XPS and high energy resolution EPES measurements were performed using the ESA-31 type electron spectrometer developed in ATOMKI [31]. In our measurements a LEG 62 (VG Microtech) type electron gun was used between 1 and 5 keV electron energies. The energy width of the primary electron beam was 0.6 eV. The scattering angle was $\theta_0 = 130°$ using an angular range $\Delta\theta_0 = \pm4°$. The angles of the incident and detected electron beams were 75° and 25° measured from the normal of the sample surface, respectively. During measurements the vacuum in the analysis chamber was better than 3×10^{-9} mbar.

The measurable spectra consist of two, more or less, separable peaks as can be seen on Fig. (**3**), where besides the total energy distributions, we also show the peaks belonging to the single scattering contribution for demonstration of the real yield ratios. In general, the absolute peak intensities cannot also be measured. Only the intensity ratio of the two peaks is measurable. Furthermore, the separation of the two peaks can be measured much more accurately than the absolute peak shifts. Therefore, in the following, we will focus on the measurable quantities.

The calculated relative energy shifts (see Table **4**), FWHMs (see Table **5**), and peak areas (see Table **6**) are in very good agreement with the experimental results.

Figs. (**5** and **6**) shows the simulated and measured energy distributions of electrons backscattered elastically from polyethylene at an incident angle of $\theta=75°$. In order to account for the instrumental effects of the analyzer and the finite line width of the electron gun in the simulated spectrum, the results of the MC calculations were convoluted by the elastic peak shape measured in the case of the Cu reference sample. Comparing the measured and simulated spectra it can be seen that our calculations are in good agreement with the experimental data not only concerning the qualitative shape of the distributions but also for the difference between the energy positions component peaks.

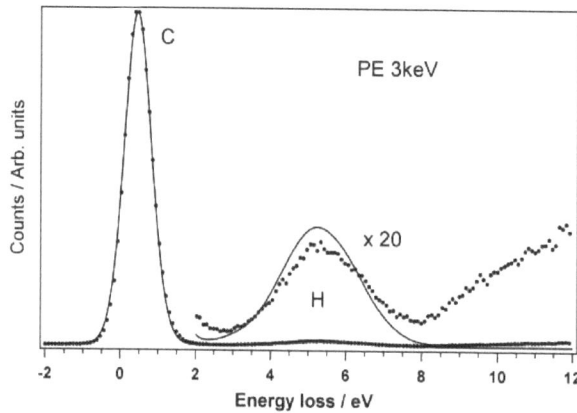

Fig. (5). The energy distribution of electrons backscattered elastically from polyethylene at $\theta = 75°$. The primary electron energy was 3 keV. Dotted line: measurement, solid line: MC simulation.

Table 4. The distance between the carbon and hydrogen peaks.

E_0 (keV)	$[E_{r0}(H) - E_{r0}(C)]$ (eV)		
	Equation (12)	Monte Carlo	Experiments
2	3.276	3.18(1)	3.13(3)
3	4.915	4.82(1)	4.88(4)
4	6.553	6.48(1)	6.47(5)
5	8.191	8.13(2)	8.20(10)

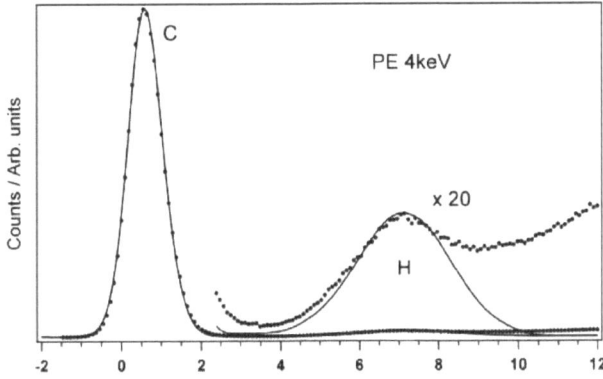

Fig. (6). The same as Fig. (**5**) but at 4 keV primary electron energy.

Table 5. The FWHM of the carbon and hydrogen peaks.

E_0	$\Delta E_r(C)(eV)$		$\Delta E_r(H)(eV)$	
(keV)	Monte Carlo	Experiments	Monte Carlo	Experiments
2	0.429(1)	0.414(2)	1.94(1)	2.26(2)
3	0.523(1)	0.532(2)	2.35(2)	2.38(4)
4	0.604(1)	0.619(2)	2.69(2)	2.91(10)
5	0.674(1)	0.685(2)	2.96(2)	3.08(13)

Table 6. The relative intensity ratios of the areas between of the hydrogen and carbon peaks as a function of the primary electron energies.

E_0	$I(H)/I(C)(\%)$	
(keV)	Monte Carlo	Experiments
2	4.96(7)	4.40(10)
3	5.04(8)	3.91(13)
4	5.05(10)	4.57(23)
5	5.03(12)	4.45(32)
2÷5	5.02±0.03	4.33±0.21

ENERGY DISTRIBUTION OF ELASTICALLY SCATTERED ELECTRONS FROM DOUBLE LAYER SAMPLES

Physical Model of the Calculations

A simple physical picture was applied in our Monte Carlo calculations and in the analytical treatment of the generation of the corresponding expressions for describing the single scattering. Our model is based on the following assumptions: a) The sample consists of two homogeneous and amorphous layers with ideally flat surfaces and interface (see Fig. 7).

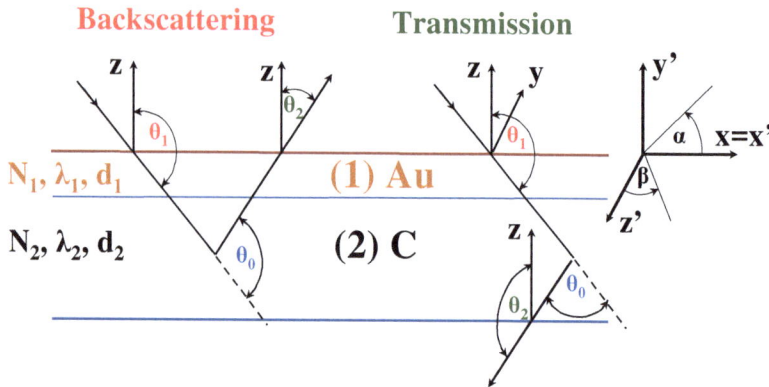

Fig. (7). Schematic view of the double layer Au-C sample and the geometry of the scattering used in our calculations. $\theta_0 = 44.3°$ was fixed in all cases.

b) The inelastic mean free path (IMFP) within the given layer is constant, and it does not change even in the vicinity of the interface. c) The surface losses in vacuum were not taken into account (which is reasonable in the case of 40 keV primary energy). d) For the description of the elastic scattering cross sections we use the calculations for free atoms, solving the Dirac-Hartree-Fock-Slater wave functions [26]. e) We suppose that the energy loss of elastically scattered electrons can be calculated within the impulse approximation, *i.e.* we neglect the final state interactions. In this case the lost energy of the electron following one elastic collision on the static target atom can be calculated as:

$$E_{r0} = \frac{2mE_0}{M_I}\left(1 + \frac{E_0}{2mc^2}\right)\left(1 - \cos\theta_0\right),\tag{14}$$

where E_0 and θ_0 are the initial energy and the scattering angle of the electron, m and M are the rest mass of the electron and the target atom, c is the velocity if the light in vacuum. The factor, $(1 + E_0/2mc^2)$ is the relativistic correction due to the velocity of the electron. For $E_0 = 40$ keV this relativistic correction causes a 3.9% increase. If we suppose that the target atom is moving, the energy loss after one elastic collision (E_r) can be written as:

$$E_r = E_{r0}\left[1 + \sqrt{\frac{M\varepsilon}{mE_0}\left(1 - \frac{E_0}{2mc^2}\right)}\frac{\cos\vartheta - \cos\theta_0\cos\vartheta - \sin\theta_0\sin\vartheta\cos\varphi}{1 - \cos\theta_0}\right],\tag{15}$$

where ε is the kinetic energy of the moving target atom, and ϑ and φ characterize the direction of the motion of the target atom with respect to the velocity of the primary electron and to the scattering plane, respectively f). We suppose that the velocity distribution of the moving target atoms in the sample is isotropic and the kinetic energy can be described by the Maxwell-Boltzmann type function:

$$P(\varepsilon)d\varepsilon = \frac{3\sqrt{3}}{\sqrt{2\pi}} \frac{\sqrt{\varepsilon}}{\bar{\varepsilon}^{3/2}} \exp\left(-\frac{2\varepsilon}{3\bar{\varepsilon}}\right)d\varepsilon, \tag{16}$$

where $\bar{\varepsilon}$ is the average kinetic energy of the atoms in the given layer. This value can be selected freely, but at room temperature, in many cases, e has values only a little higher than $3/2 \ kT$ (the temperature dependence of $\bar{\varepsilon}$ can be found in ref [32].). In the case of moving target atoms the well known Doppler effect results in a Gaussian distribution in the spectra of scattered electrons at the distance E_{r0} from the primary energy, E_0, and with ΔE_r full width at half maximum (FWHM):

$$\Delta E_r = \sqrt{8\ln 2}\sqrt{\frac{4}{3}E_{r0}\bar{\varepsilon}}\ . \tag{17}$$

We note that the Gaussian peak shape and Doppler broadening described by Eq. 17 is strictly valid only for single scattering on the identical atoms and at the same scattering angle θ_0. This condition is not usually exactly fulfilled during the real measurement, mainly due to the fact that the contribution from multiple scatterings is not negligible (or sometimes even dominant). For the case of double layers, multiple scatterings can also occur even for atoms with different masses. This type of scattering is the mixed scattering. Generally, the mixed scattering process can result in two asymmetric peaks in the energy distribution. The separation and widths of these peaks will differ smaller or larger from the values calculated according to Eqs. 15 and 17. Furthermore, experimentally the primary electron beam also has an angular divergence and the measurements can be performed only for finite solid angles, $\Delta\Omega$. If this finite solid angle can not be small enough because of the intensity purpose, then the scattering angle would not be constant, which results in different shifts and broadenings. These parameters have different weighting factors because of the changing in the differential cross section $\frac{d\sigma}{d\Omega}(\theta)$. So, the finite solid angle $\Delta\Omega$ can modify the peak shape and parameters, even in the case of the single scattering. For the case of double layer samples, the thickness of the layers also modifies the peak parameters (see later).

Therefore, in order to find the optimal experimental conditions, it is important to perform a detailed analysis of the given problem before the ERBS measurement. It would be especially useful to perform a peak shape analysis based on accurate Monte Carlo simulation, for the case of sensitive and fine experiments, like for instance, the investigation of the usefulness of the impulse approximation by the study of the final state interactions.

The Present Monte Carlo Approach

For the present simulations, our previously developed Monte Carlo code for the determination of the scattered electron spectra reflected from one and two-element samples [1, 3] was modified. The present code was capacitated to investigate the double layer samples in reflection and transmission modes (see Fig. **7**). The random motion of the electron in the sample is followed in the XYZ coordinate system. According to the usual way, random numbers describe the distances between two scattering points, the type of the scattering (elastic or inelastic), the scattering angle for the case of elastic collisions, and the direction of the further motion (θ, φ). Furthermore, in the case of elastic collisions, the energy loss is also calculated, by Eq. 15, using randomly generated angles, ϑ and φ, which describe the direction of the atomic motion in the sample, and by the help of randomly generated kinetic energy, ε, from the corresponding Maxwell-Boltzmann type energy distribution. The program does not use any other equation (for example, it does not use the expression for a peak width given by Eq. 17).

Generally, the path length between two scattering events can be calculated using a random number R, as $s = -\lambda \ln(R)$. For double layer samples, however, some modification is necessary. Our Monte Carlo program checks continuously whether the actual path length s exceeds the distance s_1, namely the distance from the given point along the scattering direction to the boarder of the two layers. If $s > s_1$ than the path length in the other layer (s_2) can be calculated as:

$$s_2 = -\lambda_2 \left[-\ln(R) - s_1 / \lambda_1 \right].$$ (18)

Eq. 18 is important for the accurate determination of the multiple scatterings, especially when the total mean free paths (λ_1, λ_2) for the two layers differ significantly. For the case of the multiple scattering when the electron was scattered elastically in both layers, the code also stored the number of scatterings separately, layer by layer. In this way, in addition to the full energy loss spectra, we can have many partial energy loss distributions.

Our code, in an earlier version, was made for the determination of the energy distributions of the scattered electrons in the entire angular range of the scattering plane (the plane which is determined by the vectors of the direction of the incident beam and the surface normal). Therefore, the solid angle, $\Delta\Omega$, was defined in the $X'Y'Z'$ coordinate system, which can be obtained from the XYZ system *via* a 90° rotation around the X axis (see Fig. **7**). In the coordinate system $X'Y'Z'$ the polar angle, $\beta = 90°\pm\Delta\beta$, is the same for all measurement channels. The $\alpha_i \pm \Delta\alpha$ azimuth angles will distinguish the individual angular channels. For illustration of the capability of the Monte Carlo code, Fig. (**8**) shows the separation of the gold and carbon peaks as a function of the emission angle for our four cases of the simulation. However, this work does not contain the quantitative evaluation of these angular distributions. In this chapter *we will focus only on one measurement channel* which corresponds to the nominal scattering angle of $\theta_0 = 44.3°$, using $\Delta\beta = 5°$ and $\Delta\alpha = 5°$, which is equivalent to a solid angle of $\Delta\Omega=0.03$ sr.

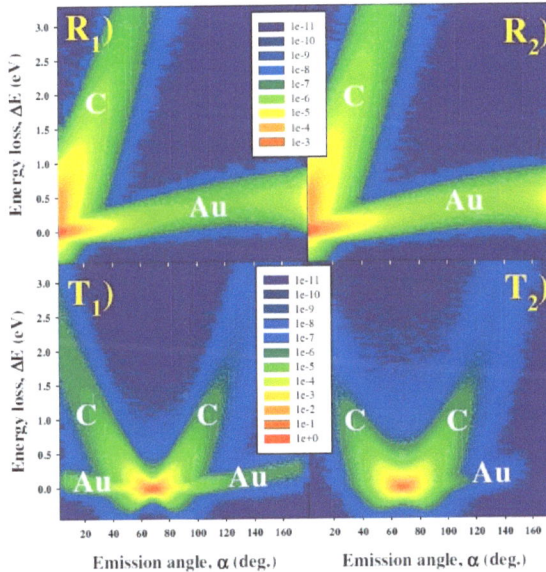

Fig. (8). Contour plot (blue: minimum intensity, red: maximum intensity) of the intensity distribution of electrons scattered or transmitted elastically from Au-C double layers at the primary electron energy of 40 keV. For the denotations see the text.

The Monte Carlo code was applied for four different cases, for two Au-C double layer samples in reflection and transmission mode (see Fig. **7**). The first sample consisted of $d_1= 1$ Å gold layer on the top of $d_2 = 90$ Å thick carbon foil and the second one consisted of $d_1= 2$ Å gold layer on the top of carbon foil with thickness of $d_2= 1400$ Å. These four cases of the simulations will be named shortly by R_1, R_2 and T_1, T_2 on the figures, where R = *reflection,* T = *transmission*, 1 and 2 denote the thinner and thicker sample, respectively. The incident angle, θ_1, was

112.15° in the reflection and 157.85° in the transmission geometry (see *XYZ* coordinate system on Fig. **7**). During the simulations these values were fixed. According to the nominal value of the scattering angle (θ_0=44.3°, which was the same in all cases), the nominal values of the emission angle (θ_2) were 67.85° in the reflection and 157.85° in the transmission geometry, respectively. Naturally, in the case of the real solid angle of the electron collection ($\Delta\Omega$) the values of the scattering and emission angles can vary in the range defined by $\Delta\alpha$=5° and $\Delta\beta$=5° (see Fig. **7**).

Our recent choice of the primary energy, incident and scattering angles was highly motivated by the experimental data published in [11, 16]. The other input data of the Monte Carlo simulations were as follows: the atomic densities were $\rho(Au)$=19.3g/cm^3 and $p(C)$=2.0g/cm^3; the corresponding inelastic mean free paths were $\lambda_i(Au)$=252.4 Å and $\lambda_i(C)$=530.3 Å; the elastic mean free paths were $\lambda_e(Au)$=48.5 Å and $\lambda_e(C)$=616.8 Å; and the total mean free paths were λ(Au)=40.7 Å and λ (C)=285.1 Å.

We used $\bar{\varepsilon}$ = 108 meV for the average kinetic energy of the C atoms. This value corresponds to the value obtained from the neutron scattering experiments [33], and with the value obtained from the earlier electron spectroscopy measurements [9], and also with our recent finding [1]. We use $\bar{\varepsilon}$ = 40 meV in the case of the gold target, because the Debye temperature is rather low for gold (165 *K),* and therefore it can not be expected to be significantly different compared to the value of 3/2*kT* [34]. We performed *theoretical experiments* and we follow 10^{11} primary electron trajectories for each collision system.

ANALYTICAL TREATMENT OF THE SINGLE SCATTERING PROBLEM

General Expressions

The main parameters of the single scattering from double layered samples, namely the scattering probabilities, peak shifts, and peak widths can be calculated analytically. In general, the single scattering probability from a layer of thickness *d* into the solid angle dΩ can be written as:

$$dP_1 = \int\limits_0^{-d/\cos\vartheta_1} \frac{ds_1}{\lambda_e} e^{-s_1/\lambda_e} \times \frac{1}{\sigma_e} \frac{d\sigma_e}{d\Omega}(\theta)d\Omega \times e^{-s_2/\lambda_e} \times e^{-(s_1+s_2)/\lambda_i} \,, \qquad (19)$$

where λ_e and λ_i are the elastic and inelastic mean free paths, respectively, σ_e is the total, and dσ_e/dΩ is the angular differential elastic cross section, s_1 is the distance from the entrance point A to the point of the first elastic scattering, and s_2 is the distance from the scattering point to the exit point (see Fig. **9**).

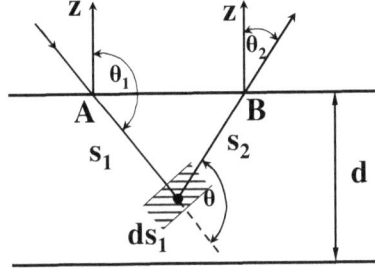

Fig. (9). Schematic diagram of the single scattering model.

The expression under the integral in Eq. 19 is built up as a product of different probabilities as follows: a) the probability of elastic scattering at the distance, s_1 and in thickness ds_1, b) the probability of the electron detection in the corresponding solid angle, c) the probability that the electron does not suffer other elastic collision in distance s_2, and finally d) that the electron does not suffer inelastic collision during the entire motion in the sample (in distance $s_{1+}s_2$). Using the relations:

$$\frac{1}{\lambda_e} = N\sigma_e, \qquad -s_1\cos\theta_1 = s_2\cos\theta_2,$$

$$0 \leq s_1 \leq \frac{-d}{\cos\theta_1}, \qquad g = \frac{\cos\theta_2}{\cos\theta_2 - \cos\theta_1}, \tag{20}$$

and

$$\frac{1}{\lambda} = \frac{1}{\lambda_e} + \frac{1}{\lambda_i}, \tag{21}$$

we can get the probability of the single elastic scattering from the first layer of the sample in the case of the reflection geometry as:

$$dP_1(1) = g\left[1 - e^{d_1/(\lambda_1 g\cos\theta_1)}\right]\lambda_1 N_1 \frac{d\sigma_1}{d\Omega}(\theta_0)d\Omega \,. \tag{22}$$

If the elastic scattering occurs in the second layer Eq. 22 is modified as:

$$dP_1(2) = ge^{d_1/(\lambda_1 g\cos\theta_1)} \left[1 - e^{d_2/(\lambda_2 g\cos\theta_1)}\right] \lambda_2 N_2 \frac{d\sigma_2}{d\Omega}(\theta_0)d\Omega , \qquad (23)$$

where N_1 and N_2 are the atomic density layer by layer and according to Fig. (7): $\cos\theta_1 < 0$ and $\cos\theta_2 > 0$. Furthermore, we note that λ_1 and λ_2, in Eqs. 22- 23 and later, denote the total scattering mean free paths calculated by the help of Eq. 21 using the elastic and the inelastic mean free paths in the corresponding layer. Similarly, to the case of the reflection mode, analytical expressions can be deduced for the transmission mode. The single scattering probability from the first layer can be written as:

$$dP_1(1) = ge^{d_2/(\lambda_2 g\cos\theta_2)} \left[e^{d_1/(\lambda_1 g\cos\theta_1)} - e^{d_1/(\lambda_1 g\cos\theta_1)}\right] \lambda_1 N_1 \frac{d\sigma_1}{d\Omega}(\theta_0)d\Omega , \qquad (24)$$

and the single scattering probability from the second layer can be read as:

$$dP_1(2) = ge^{d_1/(\lambda_1 g\cos\theta_1)} \left[e^{d_2/(\lambda_2 g\cos\theta_2)} - e^{d_2/(\lambda_2 g\cos\theta_1)}\right] \lambda_2 N_2 \frac{d\sigma_2}{d\Omega}(\theta_0)d\Omega , \qquad (25)$$

where $\cos\theta_1 < 0$ and $\cos\theta_2 < 0$ (see Fig. 7). We note that in transmission mode and for the case of $\theta_1 = \theta_2 = \theta$, the geometrical factor, g, would be infinity (see Eq. 20), however, the product of g with the expression in brackets gives $d/(\lambda\cos\theta)$. Accordingly, the yield ratio of electrons suffering only one elastic collision in layer 1 or 2 can be written as:

$$\frac{P_1(1)}{P_1(2)} = \frac{d_1 N_1 \dfrac{d\sigma_1}{d\Omega}(\theta_0)}{d_2 N_2 \dfrac{d\sigma_2}{d\Omega}(\theta_0)} . \qquad (26)$$

This ratio is independent of the mean free paths and of the geometrical factor.

Peak Shape Analysis

It is trivial that using the results of the exact analytical solutions the accuracy and validity of the Monte Carlo simulation can also be tested. Moreover, and even more importantly, the analytical calculations allow us to accurately describe the shape of the energy distribution of the electrons collected in the real solid angle $\Delta\Omega$, within the single scattering model. The energy spectra can be composed by the help of calculated recoil energies E_{r0} and elementary peak widths ΔE_r using the elementary scattering probabilities dP as weighting factors. These kind of distributions can be seen in Fig. **(10)**, for the static case ($\bar{\mathcal{E}} = 0$) and for moving carbon atoms ($\bar{\mathcal{E}} = 108$). The results of the analytical calculations are shown by solid lines and the results of Monte Carlo simulations are shown by symbols. We can observe that the corresponding spectra completely agree with each other. Fig. **(10)** shows only one example, but we obtained similar behavior in the case of 8 calculated peaks. The first moments of the energy distributions (average recoil energy), E_r, calculated by the analytical expression and by the Monte Carlo technique agree to within 0.1%. We consider this excellent agreement as an additional accurate test of our Monte Carlo calculation for the case of the single scatterings.

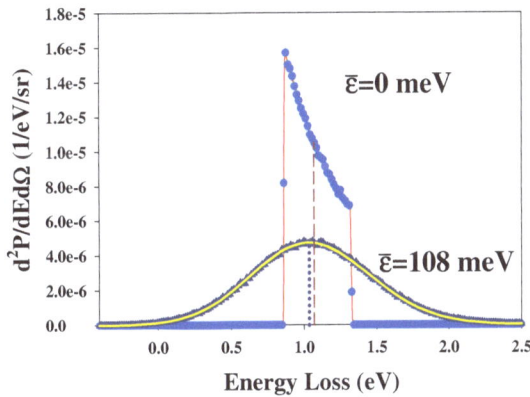

Fig. (10). Partial energy loss distribution of electrons scattered elastically from carbon atoms in Au-C double layer sample suffering only one elastic collision in reflection mode for the case of static target atom ($\bar{\mathcal{E}} = 0$ meV), and for moving carbon atoms with $\bar{\mathcal{E}} = 108$ meV. The thickness of the gold layer is 1 Å and the thickness of the carbon layer is 90 Å. The primary electron energy is 40 keV. The scattering angle is $\theta_0 = 44.3° \pm 5°$. Solid line analytical results, symbols: Monte Carlo calculations.

The shape of the curve ($\bar{\mathcal{E}} = 0$) in Fig. **(10)** is determined mainly by the change of the differential elastic scattering cross sections at all possible scattering angles in the given solid angle $\Delta\Omega$. Generally, for $E_0 = 40$ keV and in the vicinity of the scattering angle $\theta_0 = 44.3°$, the values of strongly decrease with increasing θ.

However, we note that such a sharp change of the differential cross section can also be at large scattering angles (especially in the cases of the lower primary electron energies and higher atomic numbers). The angular differential cross section, can have deep minima and thereby, depending on the scattering geometry, it can also result in such type or inverse asymmetry. The changes $\frac{d\sigma}{d\Omega}(\theta)$ in also have influence on the spectra calculated with the assumption of moving target atoms, but this effect is less conspicuous. The shape of the peak with $\bar{\varepsilon}$ = 108 meV in Fig. (**10**) is very similar to the shape of the Gaussian distribution, but not identical. They are, in fact, asymmetric distributions and the asymmetry appears in the range of the smaller (larger) energy loss.

It can be confirmed that the peak maximum is not equal to the average recoil energy $\bar{E}_r$. For our recent case, illustrated in Fig. (**10**), the peak maximum of the energy loss distribution is $E_r(max)$ = 1027 meV, while $\bar{E}_r$ = 1053 meV. The difference is 2.5%. In all cases, the peak maxima, obtained from the first derivative of the distributions, always differ from the average values of the distributions. The average deviation for our four cases was (25±0.1) meV. The right and the left half of the peaks were also fitted by different Gaussian type shapes. We found that the half width at half maximum (HWHM) for the right side was (6.6±0.1)% larger than that of the left side. These results call attention to the fact that, although the single scattering condition is fulfilled in a given experiment, it can happen that the asymmetry observed in the peak shape, or at least part of the asymmetry, is not due to the final state interaction [11], but to either the divergence of the primary beam or the applied finite solid angle, $\Delta\Omega$. Certainly, the experimentally observable peak shape contains the effect of the multiple scattering, which can be fully made out by an accurate Monte Carlo simulation. However, the main advantage of the analytical calculations is that they can be performed very accurately without any statistical errors.

Results of the Monte Carlo Simulations

We performed the Monte Carlo simulation for two Au-C bilayer samples in reflection (R_1, R_2) and transmission (T_1, T_2) mode at the vicinity of the scattering angle of θ_0=44.3° and at the primary electron energy of 40 keV. The number of elastically scattered events in the used solid angle ($\Delta\Omega$=0.03 sr) was roughly (0.5-3) x 10^6, thanks to the huge amount of primary histories (10^{11}).

The Components of the Energy Distributions

At the end of each calculation we have many partial distributions, but the most important among them are the following five: the peaks of the single scattering on Au and C atoms (see Fig. **11**), the peaks of the multiple scattering on the gold atoms only, and on the carbon atoms only (see Fig. **12**), and the energy distribution of the multiple scattered electrons when both target atoms contribute to the process (see Fig. **13**). In the first four cases we get one peak with near Gaussian shape and these peaks can unambiguously be associated either to Au or to C atoms (see Figs. **11, 12**). The fifth distribution, however, generally can consist of two peaks and a continuous spectra between the peaks, because of different masses of the scattering atoms (mixed scattering). We can evaluate this energy distribution for example with two Gaussian peaks (see Fig. **13**) and after the evaluation one peak can be associated to Au and the other one to C.

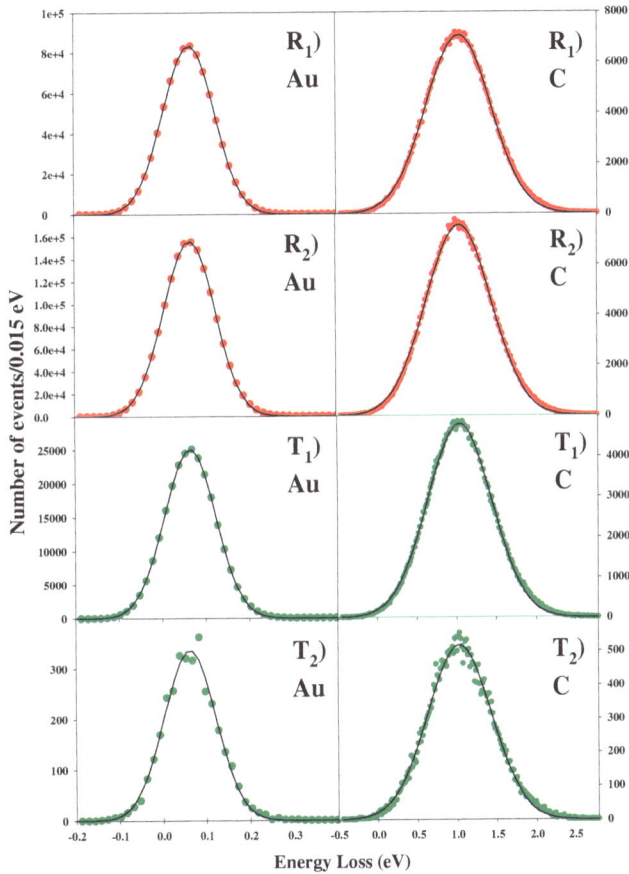

Fig. (11). Energy loss distribution of the elastically scattered electrons from Au-C double layer suffering only one elastic collision before they escape from the sample. The primary electron energy is 40 keV. Symbols: Results of our Monte Carlo simulation, solid line: Gaussian fit.

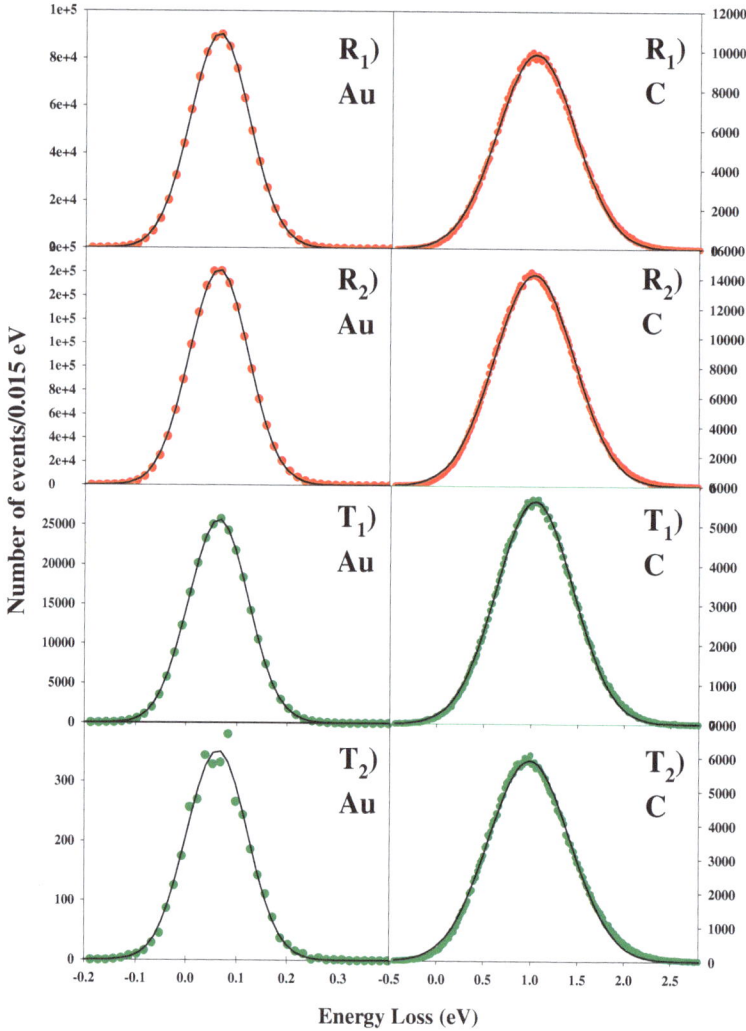

Fig. (12). Energy loss distribution of the elastically scattered electrons from Au-C double layer, suffering elastic collisions only on one type of target atoms before they escape from the sample. The primary electron energy is 40 keV. Symbols: Results of our Monte Carlo simulation, solid line: Gaussian fit.

Naturally, we have a total distribution too, which contains all electrons scattered elastically on the sample. This distribution can be compared with the experimental spectrum, if *all parameters of the measurement and simulation are exactly the same.*

The Ratios of the Peak Intensities

By integrating the energy loss distributions, we can get the relative yields (in %)

of different types of scattering in the total distributions and in the Au and C peaks separately, as shown in Table **7**. The data in parentheses (fifth column, in Table **7**) shows the values obtained from the fitting of the distribution of the mixed scattering. The other data in Table **7** belong to the direct results of the Monte Carlo simulation without any fitting procedure. From Table **7** it is clear the usefulness of the calculations, taking into account merely the contribution of the single scatterings, they are very different in these four sample - geometry combinations. About 80% of all events are single scattering for the thin sample. At the same time, in the case of the thicker sample 1/3 of the events belong to the multiple scatterings in reflection geometry, while in transmission geometry the multiple scattering is dominant (more than 90%). The data, regarding the separate peaks, show that the relative yields of the single scattering for gold and carbon are about the same in transmission geometry, but they differ considerable in reflection geometry. Generally, the multiple scattering on the carbon atoms is much more than on gold atoms, following the ratio of the thicknesses of carbon and gold layers.

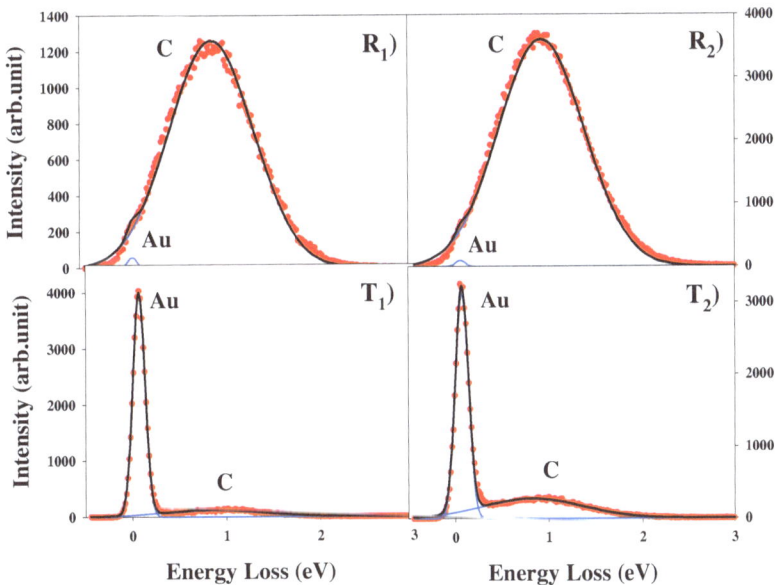

Fig. (13). Energy loss distribution of the elastically scattered electrons from Au-C double layer, suffering multiple scattering events on both target atoms before they escape from the sample (mixed scattering). The primary electron energy is 40 keV.

The contribution of the mixed scattering is roughly 6-12% of the total scattering events, but the relative yields in the gold and carbon peaks are very different. This contribution is negligible in reflection geometry and has an especially large role for thick samples in transmission geometry and in the gold peak. Accordingly, despite the fact that the primary energy of electrons *(E₀ = 40keV)*, the scattering angle (θ₀=44.3°), and the solid angle of electron collection, were the same, the three components of Au and C peaks have quite different weight factors in the different cases. These partial intensities and spectra are very useful in the data analysis and in understanding the scattering processes in double layers. Unfortunately, they can not be observed experimentally. The measurable spectra consist of two, more or less, separable peaks as can be seen on Fig. (**14**), where besides the total energy distributions we also show the peaks belonging to the single scattering contribution for the demonstration of the real yield ratios. In general, the absolute peak intensities cannot also be measured, only the intensity ratio of the two peaks is measurable. Furthermore, the separation of the two peaks can be measured much more accurately than the absolute peak shifts. Therefore, in the following, we will focus on the measurable quantities.

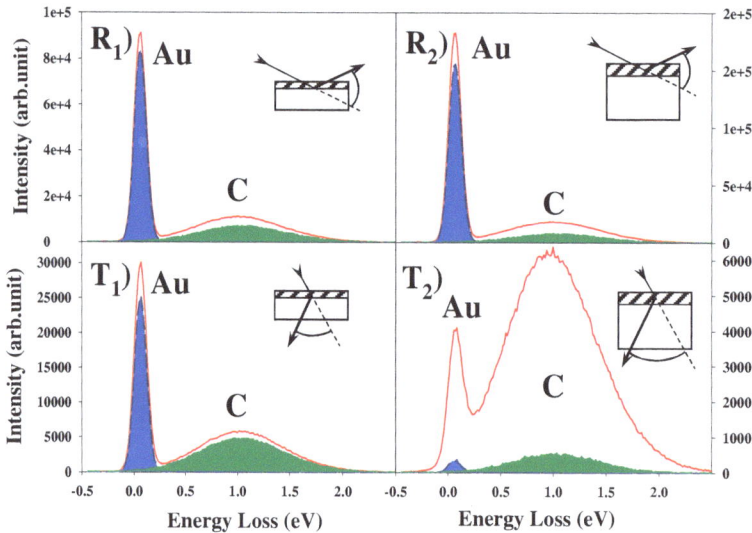

Fig. (14). Energy loss distributions of electrons scattered elastically from Au-C double layer at 40 keV primary energy for two different thicknesses in transmission and reflection geometry. Red solid line: total spectrum, blue area: the single scattering contribution from Au atoms, green area: the single scattering contribution from C atoms.

One of the aims of the ERBS measurements is the determination of the intensity ratio of the peaks of electrons scattered on the double layers.

Table **8** displays the Monte Carlo results, relative to these intensity ratios of the carbon and gold peaks as a function of the types of scattering. The second column shows the single scattering ratios at the nominal value of the scattering angle of

θ_0=44.3° according to an infinitesimally small solid angle, $d\Omega$ (Eqs. 22-25). The data in the third column shows the same ratios as the second column, but the integration was carried out for finite solid angle of $\Delta\Omega = 0.03$ sr. In the present case (E$_0$=40 keV, θ_0=$44.3°$) the intensity ratios for the single scatterings, $P_1(C)/P_1$ (Au), are very similar both for the infinitesimally small solid angles, $d\Omega$ and for the finite solid angles, $\Delta\Omega$, due to the fact that the $d\sigma/d\Omega(\theta)$ differential elastic scattering cross sections at the vicinity of the scattering angle θ_0=44.3° show similar dependence on the scattering angle for both carbon and gold. The absolute yields are 1÷4% less for the case of θ_0.

The deficit is bigger for Au in reflection mode and bigger for C in transmission mode. This can also be seen in the peak intensity ratios of C/Au, but the ratio is less sensitive for this effect. Anyhow, we can recognize that using the nominal value of the scattering angle, the C/Au intensity ratio is ~1.2% more in reflection mode and 0.6% less in transmission mode compared with the real integrated values.

We note that for smaller primary energy and larger scatterings angles one can obtain much larger differences. We consider the excellent agreement between the correct analytical calculations and the Monte Carlo results (the third and the fourth data columns) as an additional accurate test of our Monte Carlo calculation for the case of the single scatterings.

Column 5 contains the intensity ratios of the elastic scattering from different layers, since P_{mono} is the sum of the single and multiple scattering on the atoms with identical masses. These ratios show considerable differences from single scattering ratios. The values in columns 4-5 contain the pure data obtained by the Monte Carlo simulation for ratios of the partial intensities. However, for the determination of the $P(C)$ and $P(Au)$ yields containing the total number of elastic collisions, we must evaluate the spectra of mixed scatterings first.

As we mention before, this distribution contains two peaks which we fit with two Gaussian shape peaks during the evaluation. Using these peak intensities and $P_{mono}(C)$, $P_{mono}(Au)$ data we can get the C/Au intensity ratios which are denoted by the upper number in the column 6. The lower number in column 6, however, arises from the fit with the Gaussian peaks of the sum spectrum. The agreement between these two numbers is satisfactory, except for the case of the last row. In this latter case the deviation is roughly 11%, which can be attributed to the distortion effect of the multiple scatterings in the peak shape.

Table 7. Relative yields of different types of elastic electron scattering from Au-C double layers at different geometries.

[sample] (Å) Geometry	Peaks — Sum	Single scattering (%)	Multiple scattering Same atomic mass (%)	Different atomic mass (%)	$\theta_0 = 44.3°$
1 Å	Au	92.5	7.5	(≤ 0.1)	
90 Å	C	62.3	25.4	(12.3)	
Reflection	Σ	**78.42**	**15.82**	**5.76**	
2 Å	Au	86.2	13.8	(≤ 0.1)	
1400 Å	C	40.3	37.9	(21.8)	
Reflection	Σ	**66.96**	**23.91**	**9.13**	
1 Å	Au	83.8	1.9	(6.11)	
90 Å	C	82.72	14.6	(2.7)	
Transmission	Σ	**83.19**	**9.16**	**7.65**	
2 Å	Au	8.5	0.4	(91.1)	
1400 Å	C	7.7	87.0	(5.3)	
Transmission	Σ	**7.79**	**80.02**	**12.19**	

Table 8. Intensity ratios of the carbon and gold peaks in the energy spectra of electrons elastically scattered on the Au-C samples depending on the types of scattering. P_1= single scattering only, P_{mono}= single and multiple scattering on the atoms with identical masses, P= all types of elastic scatterings, including the mixed one also.

[sample] (Å) Geometry	$P_1(C)/P_1(Au)$ θ_0=44.3° (dΩ)	$P_1(C)/P_1(Au)$ Integrated (ΔΩ)	$P_1(C)/P_1(Au)$ Monte Carlo (ΔΩ)	$P_{mono}(C)/P_{mono}(Au)$ Monte Carlo (ΔΩ)	$P(C)/P(Au)$ Monte Carlo (ΔΩ)	$\dfrac{P(C)/P(Au)}{P_1(C)/P_1(Au)}$ (ΔΩ)
Au 1 / C 90 (**Reflection**)	0.5939	0.5895	0.589(1)	0.767(2)	0.875(3) 0.883(16)	1.48 1.50
Au 2 / C 1400 (**Reflection**)	0.3417	0.3365	0.337(1)	0.563(1)	0.720(2) 0.730(8)	2.14 2.17
Au 1 / C 90 (**Transmission**)	1.3071	1.3178	1.320(5)	1.517(5)	1.336(6) 1.343(10)	1.01 1.02
Au 2 / C 1400 (**Transmission**)	10.167	10.200	10.43(24)	122(2)	11.45(14) 12.75(41)	1.12 1.25

One of the most important data of our Monte Carlo simulation can be seen in the last column of Table **8**, where the intensity ratios of carbon and gold peaks from the full spectra are compared with the intensity ratios from the single scattering model. It can be seen, that while for thin sample and transmission geometry we can get surprisingly perfect agreement, for thick samples and reflection geometry the multiple scattering duplicated the $P(C)/P(Au)$ ratio compared to the single scattering one, $P_1(C)/P_1(Au)$. Since the role of the multiple scattering may strongly depend on the material constituents of the layer, the primary energy, the scattering angle, the layer thickness, the order of the layers, therefore the deviation of the measurable intensity ratio from the expected values based on the single scattering model can vary in a wide range for the double layers. For instance, if the aim of the ERBS measurement is the determination of the thicknesses of the layers, it is recommended to perform an accurate Monte Carlo simulation for the given experimental condition. It is even better to perform these calculations for the determination of the "good geometry" of the experiment. The transmission geometry with a suitable scattering angle seems to be preferable.

The Separation of the Peaks

The final energy distributions of elastically scattered electrons obtained from the Monte Carlo simulations were fitted by Gaussian peaks. As a result of the fitting, the separation between the two peaks and the full width at half maximum (FWHM) of the peaks can be determined. Table **9** shows the separation between C and Au peaks as a function of the type of the scatterings for various combination of samples and geometries.

Assuming only single scattering and taking into account the case of $E_o = 40$ keV, $\theta_0 = 44.3°$ for infinitesimally small solid angle $d\Omega$, the peak separation, $E_{r0}(C) - E_{r0}(Au)$ *is 1014 meV for every cases* (named as nominal value). According to the Table **9** we can observe two important effects. The first one is that the peak separations are smaller than the nominal values by roughly on average 3.5%, both in reflection and transmission modes, when we apply a finite solid angle, $\Delta\Omega = 0.03$ sr, mainly due to the larger weighting factor of smaller scattering angles. Secondly, if the contribution of the multiple scattering on the same atoms and the mixed scatterings are added to the contribution of the single scattering, the peak separations further decrease. Finally, the average difference from the nominal value is about 7%. The smallest difference (4.3%) is in the case of the thin sample and transmission geometry. This is in accordance with the previous observation for the intensity ratios. We also note that the change of the peak separations is dictated by the change of the C- peak shift.

Table 9. The peak separations in the energy spectra of electrons elastically scattered on the Au-C double layer samples for different types of scattering. The nominal value is 1014 meV for all cases.

[sample] (Å) Geometry	$[E_{r0}(C) - E_{r0}(Au)]$ (meV)		
	Single scattering	Mono atomic scattering	Sum of the scatterings
Au 1 / C 90 (Reflection)	984(1)	968(1)	952(1) (-6.1 %)
Au 2 / C 1400 (Reflection)	988(1)	965(1)	940(1) (-7.3 %)
Au 1 / C 90 (Transmission)	977(1)	974(1)	970(1) (-4.3 %)
Au 2 / C 1400 (Transmission)	968(3)	926(2)	910(2) (-10.3 %)

Peak Widths

Assuming only single scattering and taking into account the case of $E_0 = 40$ keV, $\bar{\varepsilon}$ (Au)=40 meV, $\bar{\varepsilon}$ (C) = 108 meV, θ_0=44.3° for infinitesimally small solid angle *dΩ, the expected values of the FWHM of Au and C peaks are 149 meV and 988 meV, respectively.* The relative changes in the shape of Au peaks are similar to the changes in the shape of C peaks, but they are much smaller and hardly observable. Therefore, in Table **10** we show only the data related to the C peaks. In the case of solid angle, $\Delta\Omega$=0.03 sr, bigger weighting factors and smaller peak widths belong to the smaller scattering angles. Therefore, for all sample and geometry combinations we found that the FWHM of the single scattering peaks are smaller than 988 meV (nominal value), but depending on the geometry, the decreases are also different. However, the effect of multiple scattering manifests itself in the broadening of the peak widths. This effect compensates completely the deficit in FWHM obtained in the single scattering peaks for the case of thick sample and transmission geometry. In other cases, it overcompensates, especially for the geometry of thick C layer in transmission mode. The overcompensation in the latter case results in the difference of +7.4% in comparison with the nominal value. Using the peak widths, evaluated from the experimentally observable spectra one can calculate the mean kinetic energy of the atoms by the simple use of Eq. 17. However, Table **10** demonstrates the effects of decreasing and increasing of the peak widths. *This means that the accurate Monte Carlo simulation can help in the evaluation of experimental data, and to get a real value of the mean kinetic energy of the atoms.*

Table 10. The widths of carbon peaks in the energy spectra of electrons elastically scattered on the Au-C double layer samples for different types of scattering. The nominal value is 988 meV for all cases.

[sample] (Å) Geometry	$\Delta E_r(C)$ (FWHM; meV)		
	Single scattering	Mono atomic scattering	Sum of the scatterings
Au 1 / C 90 (**Reflection**)	966(2)	979(2)	1006(2) (+1.8 %)
Au 2 / C 1400 (**Reflection**)	970(2)	996(2)	1034(2) (+4.7 %)
Au 1 / C 90 (**Transmission**)	959(2)	963(2)	985(3) (-0.3 %)
Au 2 / C 1400 (**Transmission**)	954(5)	1020(4)	1061(4) (+7.4 %)

Comparison with Experiments

As we have shown in the previous sections, the peak shape of the elastically scattered electron distribution scattered from bilayer sample may be influenced by many factors. These are the followings: geometry of the measurement; the primary energy and its angular dispersion; the angular dependence of the differential cross section in the given solid angle $\Delta\Omega$; the layer thicknesses; the average kinetic energy of the atoms in the sample; and the ratio between the single and multiple scatterings. Beyond these facts, the measured spectra include the instrumental contributions due to the finite energy resolution of the spectrometer and the energy distribution of the primary beam produced by the electron gun. Generally, none of them can be described by the simple Gaussian type peak shape. In the propitious case (for instance, when the asymmetry of the two contributions is contrary), however, depending on the properties of the electron gun and the spectrometer, the resultant of these two effects can be approximated by Gaussian type distribution. Moreover, the measured energy width of the elastic peak may include contributions from electrons suffering very small energy losses due to inelastic processes (as a consequence of *e.g.* phonon excitation or other excitation processes), which are inseparable from the elastic peak [24] without using extremely high energy resolution. In our present simulation we neglect these inelastic processes.

The comparison of the results of the simulation would have been correct only if *all* of the input data are exactly the same as those the corresponding experimental conditions. In the present case, the most important parameters of the simulations, like primary energy, nominal value of the scattering angle, and ingoing and outgoing angles, follow the experimental conditions of Went and coworkers [16]. However, the initial angular spread of the electron gun, the real solid angle, $\Delta\Omega$, of the measurements, the energy distribution of the primary electron beam, produced by the electron gun, the response function of the analyzer and the layer thicknesses were not known exactly. But besides these difficulties, we performed a qualitative comparison with the experimental data. In order to account for the instrumental effects of the analyzer and the finite line width of the electron gun, the results of the present Monte Carlo calculations were convoluted by Gaussian distribution. The FWHM of the Gaussian peak was selected in such a way that the FWHM of the convoluted Au peak was the same as in the measured one. Fig. (**15**) shows the measured energy distributions of electrons scattered elastically from Au-C double layer at 40 keV primary energy for two different thicknesses in transmission and reflection geometry [16] in comparison with the recent Monte Carlo simulation. According to Fig. (**15**) it can be seen that our calculations are in a good agreement with the experimental data concerning the shape of the distributions. The discrepancies can be attributed to the uncertainty in the thickness of the layers used in the measurements and in the lower value of the carbon foil density used in the simulation $(p(C) = 2g/cm^3)$.

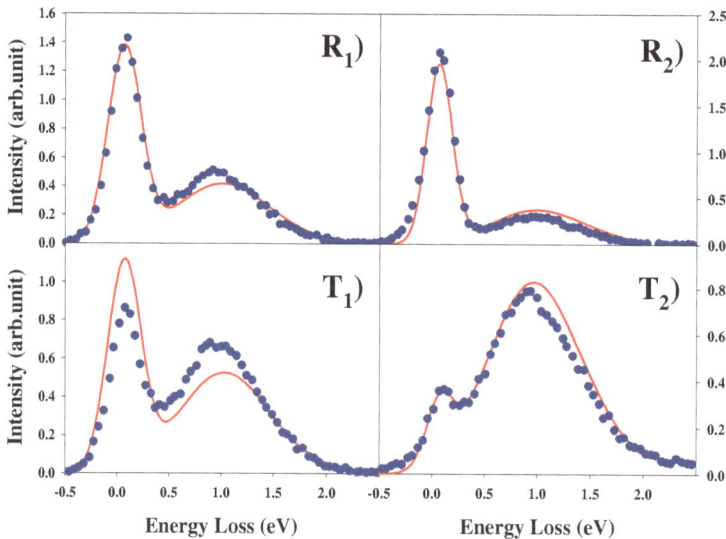

Fig. (15). Energy loss distributions of electrons scattered elastically from Au-C double layer at 40 keV primary energy for two different thicknesses in transmission and reflection geometry. Red solid line: the results of the present Monte Carlo simulation convoluted with the experimental resolution, solid circle: experimental data [16].

SUMMARY

We performed Monte Carlo simulations of electron spectra scattered elastically from multicomponent, semi-infinity sample and from double layer samples. In the first case we present detailed theoretical as well as experimental studies for the detection of hydrogen by analyzing the spectra of electrons backscattered elastically from polyethylene $((CH_2)_n)$. In the second case accurate Monte Carlo simulations were presented for two Au-C bilayer samples with various thicknesses in reflection and transmission mode. The analysis was based on very large scale Monte Carlo simulations of the recoil and Doppler effects. Besides the experimentally measurable energy distributions the simulations gave many partial distributions separately, depending on the number of elastic scatterings (single, and multiple scatterings of different types). We also present detailed analytical calculations for the main parameters of the single scattering, taking into account both the ideal scattering geometry, *i.e.* infinitesimally small angular range, and the effect of the real, finite angular range used in the measurements. The effect of the multiple scattering on intensity ratios, peak shifts and broadening, were shown.

We calculated the yields and energy loss distributions of electrons backscattered elastically from polyethylene sample at $\theta = 130° \pm 5°$ and in the primary energy range between 1 and 5 keV using the Monte Carlo technique. Besides the total yield and spectrum, the yields and spectra for single, mono-atomic and mixed scattering were also calculated. The corresponding data takes into account only single scattering events were also obtained analytically. These data were used to test end calibrate our Monte Carlo code.

The well separated H elastic peak can be easily utilized for estimating relative hydrogen content of materials with similar composition, comparing the respective peak intensities. However, quantitative determination of hydrogen concentration in the case of multicomponent samples is not simple. First of all, escape of H atoms during irradiation of the sample by electrons in vacuum, should be avoided as much as possible, optimizing the primary beam current. In the case of insulator samples, effects of surface charging should be minimized. Because of the low relative intensity of the hydrogen elastic peak the proper modeling of the inelastic background in the spectra has an increased importance concerning the accuracy of quantitative analysis. In addition, for the interpretation of the experimental spectra a correct Monte-Carlo simulation, based on realistic surface concentrations of the other components (*e.g.* derived from XPS measurements), is necessary. This is confirmed by our results indicating a ca. 40% contribution to the H elastic peak from "mixed" multiple elastic scattering processes involving different type of atoms.

We also found that the experimentally obtained total energy loss distribution consists of two separate peaks and between the two peaks a low intensity continual electron distribution is located. Although the two peaks can be well described with Gaussian peaks, the distance between the two peaks is smaller and those FWHMs are greater than the values calculated from the single scattering model. The peak intensity ratios (area of the peaks), with increasing primary energy, show sign exchange (from positive to negative) compared with the ratios obtained from the single scattering model. These deviations are caused by the so called mixed scatterings when the scatterings are of both types of atoms in the sample. The deviations even for the same scattering geometry and the sample depend on the primary energy.

Our calculations show that during the evaluation (peak and background fitting) of the experimentally obtained spectra of the electrons backscattered elastically from the semi-infinity sample with high energy resolution, and during the deductions from the evaluated data (concentration, average kinetic energy of atoms), the effect of the mixed scatterings has to take into account. These effects can be analyzed by Monte Carlo method.

We calculated the yields and energy loss distributions of electrons backscattered elastically from two Au-C bilayer samples with various thicknesses in reflection and transmission mode taking into account both the recoil effect and Doppler broadening due to the atomic motion.

We used two partial distributions, namely the distributions of the single scattering on gold and carbon atoms, for the test of our Monte Carlo code. The results of our Monte Carlo calculations for yields, peak shifts and peak widths were compared with detailed analytical calculations. We found that the corresponding data are in excellent agreement with each other to within 0.1%.

By the help of three further partial distributions (two distributions of the multiple scattering on the gold atoms only, and on the carbon atoms only, and finally the energy distribution of the multiple scattered electrons when both target atoms contribute to the process) the effect of the multiple scatterings on the intensity ratios of the gold and carbon peaks, the peaks shift, and the peak widths were investigated.

From the evaluation of the summed energy loss spectra we found the followings: a) While for transmission geometries, especially for the case of thin sample, we found good agreement with the results of the single scattering model in the peak intensity ratios of gold and carbon, particularly large differences (50-100%) were obtained in reflection geometries. b) In our recent cases, the separation of the gold and carbon peaks is 6-10% smaller than the nominal values, calculated from the primary energy and scattering angle (and mean kinetic energies), due to the finite solid angle used ($\pm 5°$), the strongly decreasing angular differential cross sections in the range of the solid angle, and the multiple scatterings. c) The widths of the carbon peaks in the summed spectra are 0-7% larger than can be expected from the nominal value. This increase is due to the fact that the effect of the multiple scattering overcompensates the effect of the final solid angle which causes a (2-3)% decrease. These results are based on the fit of the peaks by Gaussian peak. This fit seems to be a reasonable approximation; we note, however, that the peaks are asymmetric even for the case of the single scattering due to the finite solid angle used.

Other important result of our present work is with regard to the combination of analytical investigation with numerical integration, using the single scattering collision model. The accurate elastic scattering probabilities and energy distributions were calculated for gold and carbon atoms in transmission and reflection mode also taking into account the finite solid angles by numerical integration. The peaks in the energy distributions can be fitted reasonable well by Gaussian peaks. However, in all cases the peaks were asymmetric at the higher energy loss side of the spectra. The right and the left half peaks were also fitted by Gaussian type peaks. We found that the HWHM for the right side was (6.6±0.1)% larger than that of the left side. It is important that this asymmetry can exist in the frame of the impulse approximation (in absence of the final state interactions) due to the change of the scattering angle, θ, and differential cross sections, respectively. We note, however, that for the interpretation of the experimental data by the analytical calculation of the single scattering in many cases is not enough, especially for reflection geometry and at a primary energy as high as 40 keV. It is practical to use a Monte Carlo simulation where the effects of the multiple scattering are taken into account. Finally, we close by pointing out that although the calculations were performed at fixed energy and scattering angles, the description developed in this chapter can be also applied to a wide range of incident energies and for a wide range of scattering angles.

CONSENT FOR PUBLICATION

Not applicable.

CONFLICT OF INTEREST

The author (editor) declares no conflict of interest, financial or otherwise.

ACKNOWLEDGMENT

This work was supported by the National Research, Development and Innovation

Office (NKFIH) grant No. KH126886.

REFERENCES

[1]   D. Varga, K. Tőkési, Z. Berényi, J. Tóth, L. Kövér, G. Gergely, and A. Sulyok, "Energy shift and broadening of the spectra of electrons backscattered elastically from solid surfaces", *Surf. Interface Anal.,* vol. 31, pp. 1019-1026, 2001.
        [http://dx.doi.org/10.1002/sia.1121]

[2]   F. Yubero, V.J. Rico, J.P. Espins, J. Cotrino, and A.R. Gonzalez-Elipe, "Quantification of the Hydrogen content in DLC and polymeric thin films by REELS", *Appl. Phys. Lett.,* vol. 87, p. 084101, 2005.
        [http://dx.doi.org/10.1063/1.2011786]

[3]   D. Varga, K. Tőkési, K. Berényi, J. Tóth, and L. Kövér, "Observation of the hydrogen peak in the spectra of electrons backscattered from polyethylene", *Surf. Interface Anal.,* vol. 38, pp. 544-547, 2006.
        [http://dx.doi.org/10.1002/sia.2231]

[4]   V.J. Rico, J.P. Espins, J. Cotrino, A.R. Gonzalez-Elipe, D. Garg, and S. Henry, "Determination of the hydrogen content in diamond-like carbon and polymeric thin films by reflection electron energy loss spectroscopy", *Diamond Related Materials,* vol. 16, pp. 107-111, 2007.
        [http://dx.doi.org/10.1016/j.diamond.2006.04.002]

[5]   B. Lesiak, J. Zemek, and J. Houdkova, "Hydrogen detection and quantification at polymer surfaces investigated by elastic peak electron spectroscopy (EPES)", *Polymer (Guildf.),* vol. 49, pp. 4127-4132, 2008.
        [http://dx.doi.org/10.1016/j.polymer.2008.07.029]

[6]   M. Filippi, and L. Calliari, "On the use of elastic peak electron spectroscopy (EPES) to measure the H content of hydrogenated amorphous carbon films", *Surf. Interface Anal.,* vol. 40, pp. 1469-1474, 2008.
        [http://dx.doi.org/10.1002/sia.2932]

[7]   H. Boersch, R. Wolter, and H.Z. Schoenebeck, "Elastische Energieverluste kristallgestreuter Elektronen", *Z. Phys.,* vol. 199, pp. 124-134, 1967.
        [http://dx.doi.org/10.1007/BF01326021]

[8]   M. Vos, "Observing atom motion by electron-atom Compton scattering", *Phys. Rev.,* vol. A 65, p. 012703, 2001.
        [http://dx.doi.org/10.1103/PhysRevA.65.012703]

[9]   M. Vos, "Detection of hydrogen by electron Rutherford backscattering", *Ultramicroscopy,* vol. 92, no. 3-4, pp. 143-149, 2002.
        [http://dx.doi.org/10.1016/S0304-3991(02)00127-4] [PMID: 12213015]

[10]   M.R. Went, and M. Vos, "High-resolution study of quasi-elastic electron scattering from a two-layer system", *Surf. Sci.,* vol. 600, pp. 2070-2078, 2006.
        [http://dx.doi.org/10.1016/j.susc.2006.02.038]

[11]   M. Vos, and M. R. Went, "Effects of bonding on the energy distribution of electrons scattered elastically at high momentum transfer", *Phys. Rev.,* vol. B 74, p. 205407, 2006.
        [http://dx.doi.org/10.1103/PhysRevB.74.205407]

[12]   M.R. Went, and M. Vos, "Investigation of binary compounds using electron Rutherford backscattering", *Appl. Phys. Lett.,* vol. 90, p. 072104, 2007.
        [http://dx.doi.org/10.1063/1.2535986]

[13]   M. Vos, and M.R. Went, "Experimental confirmation of the EPES sampling depth paradox for overlayer/substrate systems", *Surf. Sci.,* vol. 601, pp. 1536-1543, 2007.
        [http://dx.doi.org/10.1016/j.susc.2007.01.014]

[14]   M. Vos, and M.R.J. Went, "Elastic electron scattering at high momentum transfer: A possible new analytic tool Electron Spectrosc", *Relat. Phenom.*, vol. 155, pp. 35-39, 2007. [http://dx.doi.org/10.1016/j.elspec.2006.09.003]

[15]   M.R. Went, "Vos, MElectron Rutherford back-scattering case study: oxidation and ion implantation of aluminium foil", *Surf. Interface Anal.*, vol. 39, pp. 871-876, 2007. [http://dx.doi.org/10.1002/sia.2603]

[16]   M.R. Went, M. Vos, and R.G. Elliman, "Electron inelastic mean free path in solids as determined by electron Rutherford back-scattering", *J. Electron Spectrosc. Relat. Phenom.*, vol. 156-158, pp. 387-392, 2007. [http://dx.doi.org/10.1016/j.elspec.2006.11.041]

[17]   M. R. Went, and M. Vos, "Rutherford backscattering using electrons as projectiles: Underlying principles and possible applications", *Nucl. Instr. and Meth.*, vol. B 266, pp. 998-1011, 2008.

[18]   Z.J. Ding, Y.G. Li, S.F. Mao, R.G. Zeng, H.M. Li, Z.M. Zhang, and K. Tőkési, "Monte Carlo calculation of electron Rutherford backscattering spectra and high-energy reflection electron energy loss spectra", *Nucl. Instrum. Methods Phys. Res.*, vol. B267, pp. 215-220, 2009.

[19]   C.A. Chatzidimitriou-Dreismann, M. Vos, C. Kleiner, and T. Abdul-Redah, "Comparison of electron and neutron compton scattering from entangled protons in a solid polymer", *Phys. Rev. Lett.*, vol. 91, no. 5, p. 057403, 2003. [http://dx.doi.org/10.1103/PhysRevLett.91.057403] [PMID: 12906632]

[20]   F. Yubero, and K. Tőkési, "Appl. Identification of hydrogen and deuterium at the surface of water ice by reflection electron energy loss spectroscopy", *Phys. Lett.*, vol. 95, p. 084101, 2009.

[21]   C.M. Kwei, Y.C. Li, and C.J. Tung, "Energy spectra of electrons quasi-elastically backscattered from solid surfaces", *J. Phys. D Appl. Phys.*, vol. 37, pp. 1394-1400, 2004. [http://dx.doi.org/10.1088/0022-3727/37/9/014]

[22]   A. Jablonski, P. Mrozek, G. Gergely, M. Menyhárd, and A. Sulyok, "The inelastic mean free path of electrons in some semiconductor compounds and metals", *Surf. Interface Anal.*, vol. 6, pp. 291-294, 1984. [http://dx.doi.org/10.1002/sia.740060609]

[23]   A. Jablonski, P. Mrozek, G. Gergely, M. Menyhárd, and A. Sulyok, "The inelastic mean free path of electrons in some semiconductor compounds and metals", *Surf. Interface Anal.*, vol. 6, pp. 291-294, 1984. [http://dx.doi.org/10.1002/sia.740060609]

[24]   G. Gergely, M. Menyhárd, Zs. Benedek, A. Sulyok, L. Kövér, J. Tóth, D. Varga, Z. Berényi, and K. Tőkési, "Recoil broadening of the elastic peak in electron spectroscopy", *Vacuum*, vol. 61, pp. 107-111, 2001. [http://dx.doi.org/10.1016/S0042-207X(00)00464-4]

[25]   K. Tőkési, D. Varga, L. Kövér, and T. Mukoyama, "Monte Carlo modelling of the backscattered electron spectra of silver at the 200 eV and 2 keV primary electron energies", *J. Electron Spectrosc. Relat. Phenom.*, vol. 76, pp. 427-432, 1995. [http://dx.doi.org/10.1016/0368-2048(96)80006-5]

[26]   F. Salvat, and R. Mayol, "Elastic scattering of electrons and positrons by atoms. Schrödinger and Dirac partial wave analysis", *Comput. Phys. Commun.*, vol. 74, pp. 358-374, 1993. [http://dx.doi.org/10.1016/0010-4655(93)90019-9]

[27]   S. Tanuma, C.J. Powell, and D.R. Penn, "Calculations of electron inelastic mean free paths. V. Data for 14 organic compounds over the 50–2000 eV range", *Surf. Interface Anal.*, vol. 21, pp. 165-176, 1994. [http://dx.doi.org/10.1002/sia.740210302]

[28]   K. Tőkési, and D. Varga, "Energy distributions in quasi-elastic peak electron spectroscopy", *Surf. Sci.*,

vol. 604, pp. 623-626, 2010.
[http://dx.doi.org/10.1016/j.susc.2010.01.005]

[29]   D. Varga, L. Kövér, J. Tóth, K. Tőkési, B. Lesiak, A. Jablonski, C. Robert, B. Gruzza, and L. Bideux, "Determination of yield ratios of elastically backscattered electrons for deriving inelastic mean free paths in solids", *Surf. Interface Anal.,* vol. 30, pp. 202-206, 2000.
[http://dx.doi.org/10.1002/1096-9918(200008)30:1<202::AID-SIA798>3.0.CO;2-9]

[30]   Z. Berényi, B. Aszalós-Kiss, J. Tóth, D. Varga, L. Kövér, K. Tőkési, I. Cserny, and S. Tanuma, "Inelastic mean free paths of Ge in the range of 2–10 keV electron energy", *Surf. Sci.,* vol. 566, pp. 1174-1178, 2004.
[http://dx.doi.org/10.1016/j.susc.2004.06.080]

[31]   L. Kövér, D. Varga, I. Cserny, J. Tóth, and K. Tőkési, "Some Applications of high-energy, high-resolution Auger electron spectroscopy using bremsstrahlung radiation", *Surf. Interface Anal.,* vol. 19, pp. 9-15, 1992.
[http://dx.doi.org/10.1002/sia.740190106]

[32]   "Fujikawa; Suzuki, R.; Kövér, L. Theory of recoil effects of elastically scattered electrons and of photoelectrons", *J. Elect. Spect. Relat. Phenom.,* vol. 151, pp. 170-177, 2006.
[http://dx.doi.org/10.1016/j.elspec.2005.11.011]

[33]   J. Mayers, T.M. Burke, and R.J. Newport, "Neutron Compton scattering from amorphous hydrogenated carbon", *J. Phys. Condens. Matter,* vol. 6, pp. 641-659, 1994.
[http://dx.doi.org/10.1088/0953-8984/6/3/006]

[34]   C. Kittel, *Introduction to solid state physics.* 7th. Wiley: New York, 1996.

CHAPTER 8

Four-body Effects in the ^{6}He + ^{58}Ni Scattering

Viviane Morcelle[1], **Manuela Rodríguez-Gallardo**[2] and **Rubens Lichtenthäler**[3,*]

[1] Departamento de Física, Universidade Federal Rural do Rio de Janeiro, CEP. 23890-000, Rio de Janeiro, Brazil

[2] Departamento de Física Atómica, Molecular y Nuclear, Universidad de Sevilla, Apdo. 1065, E-41080, Sevilla, Spain

[3] Instituto de Física da Universidade de São Paulo, Depto. de Física Nuclear, CEP. 05508-090, São Paulo, Brazil

Abstract: Most of the knowledge of the atomic nucleus was obtained from experimental data involving stable nuclei or nuclei in the vicinity of the stability line. Since the 1980s, several intermediate energy laboratories in the world started to produce nuclei out of the stability line, Rare Ion Beams (RIB). Many new interesting phenomena related to these nuclei have been discovered so far. Light nuclei far away from the stability line such as ^{6,8}He, ^{11}Be, ^{11}Li, ^{22}C, ^{24}O and others have been produced in laboratory. Some of these nuclei present a pronounced cluster structure formed by a core plus one or more loosely bound neutrons forming a kind of low density nuclear matter around the core (nuclear halo). Most of the research involving RIB was developed at intermediate energies, from 30 up to hundreds of MeV/nucleon, and more recently, some facilities are producing secondary beams to perform scattering experiments at energies around the Coulomb barrier. Heavy ion elastic scattering angular distributions at incident energies close to the Coulomb barrier, when plotted as a ratio to the Rutherford cross section, frequently exhibit a typical Fresnel type diffraction pattern, with oscillations around $\sigma/\sigma_{Ruth} \sim 1$ in the forward angle region, followed by a strong peak and a subsequent fall of the ratio σ/σ_{Ruth} at backward angles. This behaviour is a consequence of the interference between the Coulomb and nuclear scattering amplitudes. Due to the low binding energies of exotic projectiles, the coupling between the elastic channel and the breakup states of the projectile is very important and strongly affects the elastic angular distributions, with a damping of the Fresnel oscillations and the complete disappearance of the Fresnel peak in some cases. To describe the effect of the breakup of the projectile in the elastic scattering, new theoretical approaches have been developed. We present ^{6}He + ^{58}Ni elastic scattering angular distributions measured at three energies a little above the Coulomb barrier. The angular distributions have been analyzed by Continuum-Discretized Coupled-Channels calculations to take into account the effect of the ^{6}He breakup on the elastic scattering.

---

* Corresponding author Rubens Lichtenthäler: Instituto de Física da Universidade de São Paulo, Depto. de Física Nuclear, CEP. 05508-090, São Paulo, Brazil; Tel: + 55 11 3091 6847; E-mail: rubens@if.usp.br

Antônio Carlos Fontes dos Santos (Ed.)

Two different approaches were used to describe the structure of the projectile. One considering the ^{6}He as a three-body system consisting of an alpha particle and 2 neutrons which, in addition to the target, form a four-body problem. To compare, in a second approach, the projectile is described as a two-body cluster formed by an alpha particle plus a di-neutron. A new kind of effect due to the projectile breakup in the elastic scattering angular distributions has been reported.

Keywords: Nuclei, Nuclides, Oscillations, Projectile.

1. INTRODUCTION

The atomic nucleus is a quantum many body system consisted of protons and neutrons which interact *via* the strong nuclear and Coulomb forces. Due to the fact that the strong nuclear force is still not completely understood and owing to the complexity of the many body problem, a complete theoretical description of the nucleus is still missing. The development of nuclear physics over the 20th century was done in close connection with experiment and, most of the knowledge obtained from this research was based on the study of stable nuclei or nuclei nearby the stability line. Since the 1980s, Rare Ion Beam (RIB) facilities entered into operation all around the world allowing the production of secondary beams of nuclei out of the stability line. As a result, the number of known nuclides increased from approximately 1200 in the 1960's up to about 3000 nuclides nowadays, revealing a huge field of research to be explored. With the advent of RIB facilities, the quest in the nuclear chart has been extended from one-dimensional, where only the mass of nucleus can be varied, to a two-dimensional one, where both the number of protons and neutrons can be varied. As a consequence, many new questions emerged. The first question is the location of the proton and neutron drip lines, *i.e.*, the limits of the nuclear existence beyond which no bound nuclei exist. A remarkable finding from the research in this area concerns the nuclear shell structure, which can apparently be modified as we approach to the drip lines. The disappearance of known magic numbers and the appearance of new ones for nuclei near to the neutron drip line has been reported [1, 2].

The presence of nuclei out of the stability line in stars could have consequences for nuclear astrophysics. The existence of unstable nuclei with sufficiently long half-lives could, in principle, affects the nucleosynthesis in explosive scenarios, such as type II supernovae. The r-process passes through intermediate mass neutron rich nuclei located close to the neutron drip line to produce elements heavier than iron. On the other hand, light exotic nuclei such as ^{6}He and ^{8}Li could act as bridges to overcome the $A=5$, 8 mass gaps to the synthesis of heavier elements (see for example Ref [3].).

One of the most striking results of the RIB research was the synthesis of ^{11}Li by Tanihata in 1985. The nucleus ^{11}Li has an unusual structure formed by a ^{9}Li core plus two weakly bound neutrons. Due to their low binding energies and low angular momentum, the neutron wave functions tunnel through the barrier, extending over large distances from the core. As a consequence, a kind of low density neutron halo is formed around the ^{9}Li core of ^{11}Li, with a radius much larger than its partners ^{6,7,9}Li [4].

Other neutron rich nuclei such as ^{6,8}He, ^{11,12,14}Be, ^{22}C have similar structure with a core plus one or more weakly bound neutrons. These so-called exotic nuclei, present neutron separation energies of a few hundreds of keV, much smaller than the usual binding energies of 7-10 MeV found along the stability line. Their half-lives of few milliseconds up to hundreds of milliseconds are, in many cases, sufficiently large to allow the production of secondary beams of these elements in laboratory.

In addition, nuclei such as ^{6}He (alpha+n+n) and ^{11}Li (^{9}Li+n+n) present a three-body structure (core+n+n) since they are bound only if the three particles are present. Any two-body sub-system formed by core+n or n+n is unbound.

Nuclei presenting this three-body structure are called Borromean nuclei, and can be considered as 'laboratories' for the study of three-body forces. Three-body forces seem to be an essential ingredient for the theoretical description of the nuclear many body problem and there are not many examples of Borromean type nuclei along the stability line. Nuclei such as ^{9}Be and ^{12}C are formed respectively by two alpha particles plus one neutron and three alpha particles and are also Borromean since they are bound only if the three particles are present. However the exotic ^{6}He and ^{11}Li have the peculiarity of being composed of particles which do not interact *via* the Coulomb force what simplifies the theoretical description of the three-body problem. Nuclei with these properties may have importance in the research of more general issues such as in the many-body problem in physics and the possible existence of Efimov states [5, 6].

Most of the research with secondary exotic beams has been carried out over the last decades at intermediate energy facilities, where the energies of the secondary beams range from 30 up to hundreds of MeV/nucleon. More recently, some laboratories in the world are dedicated to produce low energy secondary beams of exotic nuclei. This is the case of the Radioactive Ion Beam in Brazil (RIBRAS) facility, which operates since 2004. Secondary beams of ^{6}He, ^{8}Li, ^{7}Be, ^{8}B and other light elements have been produced in RIBRAS and scattering experiments have been performed, using combinations of the available exotic beams and different target masses. A systematic study of the total reaction cross sections of systems with exotic projectiles, at energies around and above the Coulomb barrier has been performed.

In particular, the ^{6}He is a Borromean neutron halo nucleus formed by an alpha particle plus two neutrons bound by 0.973 MeV. Due to the neutron halo and its low binding energy, it is expected that the ^{6}He breakup process plays an important role in the elastic scattering angular distributions. The effect of the projectile breakup has been observed previously in elastic scattering angular distributions of ^{6}He, ^{11}Be and ^{11}Li on heavy mass targets such as ^{209}Bi, ^{208}Pb and ^{120}Sn [7 - 12]. A strong damping in the Fresnel peak, observed in the forward angles scattering on stable heavy nuclei, seems to be a characteristic of the angular distributions of ^{6}He, ^{11}Be and ^{11}Li on heavy targets.

The theoretical description of this phenomena can be provided by the Continuum-Discretized Coupled-Channels (CDCC) formalism [13, 14]. Because of their weak binding, halo nuclei are readily excited to the unbound states by the differential forces exerted on the constituents through the nuclear and Coulomb interactions with a target. Theoretically, explicit treatment of these breakup channels is difficult, as the states involved form a continuum of energies and are not square integrable. Hence, a robust discretization and truncation scheme, to provide a finite and normalizable basis to represent the continuum, is required. In the literature there are many methods available to discretize the continuum. Within the CDCC formalism, the binning procedure has been the method traditionally used for two-body projectiles [13, 14] and it has been extended more recently for three-body projectiles [15]. Another method that has been extensively used within the CDCC formalism, for two- and three-body projectiles, is the Pseudo-State method [16 - 21], which consists of representing the continuum spectrum of the projectile by the eigenstates of its internal Hamiltonian in a basis of square-integrable functions.

In the CDCC framework, the coupling between the elastic scattering and the projectile breakup channel is taken into account explicitly. First developed for three-body systems (two-body projectile plus a target), it was later extended to four-body systems (three-body projectile plus a target) [15, 18, 19]. It has also been developed for the inclusion of the core excitations of the projectile [22]. The formalism has been applied successfully to many reactions during last decades.

In this contribution we present results of the ^{6}He + ^{58}Ni elastic scattering measured at three energies, 12.2, 16.5, 21.7 MeV, slightly above the Coulomb barrier V_{cb}=8.7 MeV in the laboratory system [23]. To analyze the data we use the four-body CDCC formalism together with the binning procedure to represent the states of the projectile ^{6}He, treated here as a three-body system [15]. These results are compared with three-body CDCC calculations that uses a dineutron model for ^{6}He. The dineutron model can be the standard, taking for ^{6}He the experimental binding energy, or the improved model, taking an effective two-neutron separation energy in ^{6}He that reproduces the neutron density given by a three-body model [24].

In section 2, we present an overview of the experimental methods used to produce RIB and a description of the RIBRAS facility and of the ^{6}He + ^{58}Ni experiment. In section 3 we describe the CDCC formalism used to analyze the data. In section 4 an analysis of the total reaction cross sections is presented and finally in section 5 we draw the conclusions.

2. EXPERIMENTAL METHOD AND RESULTS

Nuclei out of the stability line are produced in nuclear reactions involving stable projectiles and targets. The cross sections for the production of nuclei out of the stability line increase with the bombarding energy and get to values around hundreds of millibarns in multifragmentation reactions at energies above 30 MeV/nucleon. For this reason most of the facilities for RIB production in the World operate in the intermediate energy region. There are two methods used to produce secondary beams. The in-flight and the On-lin Isotopic Separation (ISOL) methods. In the first case, a stable beam produced by a driver accelerator impinges on a thin target where the projectile fragmentation occurs.

Secondary particles are produced in a forward angle cone, with speeds near to the speed of the incoming beam. Devices such as dipoles, quadrupoles, solenoids and velocity filters can be used downstream to select the secondary beam of interest. Combinations of such devices can be used to purify the secondary beam up to a high degree of purity although it is very difficult to obtain a hundred per cent pure secondary beam using the in-flight method. In addition, the energy resolution and emittance of the secondary beam are related to the momentum transfer process in the primary target and, in general, are much worse than energy resolution and emittance of primary beams.

The ISOL method uses a primary target sufficiently thick to completely stop the primary beam. Since the primary target becomes heated by the primary beam power, the produced particles diffuse by Brownian movement through the target matter up to an ion source where they are ionized and re-injected in a secondary accelerator. In this method the secondary beam is produced from thermal energies and has properties, such as emmitance and energy resolution, of the same quality of a primary beam. Despite the better quality of the secondary beams obtained by the ISOL method, the intensities are in general smaller than the ones by the in-flight method due to the fact that the diffusion, effusion, ionization and the reaccelerating processes take time and imply in losses.

The ISOL method has been used in many intermediate energy facilities all over the World. The high energy of the primary beam is an important component since the production yield by fragmentation is directly proportional to the range of the primary beam particles in the primary target matter.

The intensities of the secondary beams produced in the first generation RIB facilities are in the range from a few particles per second up to 10^7 pps depending on the beam. These numbers are a factor of 10^{-6} smaller than the primary beam intensities, which are of the order of 10^{12} pps or more. Second generation facilities are entering into operation nowadays, and promise secondary beam intensities of 10^{9-11}, achieving intensities comparable to the primary beam intensities.

At low energies, the production of nuclei away from the stability line becomes more difficult. The production cross sections are much smaller, of the order of tens of millibarns for usual nucleon transfer and fusion reactions. No fragmentation reactions occur below 10 MeV/nucleon and the ISOL method can be applied only using production reactions such as fission of heavy elements, which have high cross sections, but produce elements in a limited region of masses.

For the production of light exotic at low energies, the in-flight method is more suitable. However, the selectors must have a high acceptance in order to collect most of the particles produced in the primary target, to compensate the reduced production cross sections. Solenoids fulfill this need, with acceptances of the order of 30 msr, a factor six larger than the acceptances of normal dipoles.

2.1. The Ribras Facility

The Radioactive Ion Beam (RIBRAS) facility operates since 2004 [25] in connection with the 8MV Pelletron tandem accelerator of the Institute of Physics of the University of São Paulo, Brazil. It consists of two superconducting solenoid in line with maximum field of *6.5* Tesla each. The scheme below shows the main components of this system (see Fig. **1**).

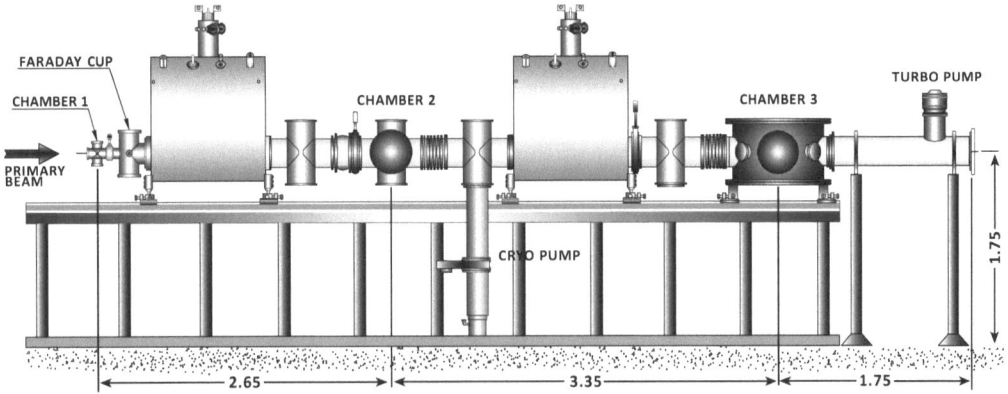

Fig. (1). RIBRAS scheme. Distances are in meters.

The secondary particles are produced by reactions between the primary beam and the primary target located in the scattering chamber 1. The primary beam is collected by a Faraday cup placed in an ISO-chamber just after the primary target. The Faraday cup has the double purpose of suppressing the primary beam and measuring its current. The number of primary beam particles impinging the primary target can be obtained by integrating the primary beam current.

The usual production reactions, secondary beams, intensities and resolutions produced so far are shown in Table **1** [26].

Table 1. Secondary beams produced at RIBRAS with production reactions, intensities, energy resolution/energy and purity measured in the scattering chamber 2.

Secondary beam	Production reaction	Q-value (MeV)	Intensity (pps)	Energy resolution FWHM (keV)/Energy	Purity %
^{6}He	^{9}Be(^{7}Li,^{6}He)^{10}B	-3.390	10^5 -10^6	1000/22 MeV	16
^{7}Be	^{3}He(^{6}Li,^{7}Be)d	+0.112	10^4 -10^5	800 / 18.8 MeV	2
^{7}Be	^{7}Li(^{6}Li,^{7}Be)^{6}He	-4.369	10^4 -10^5	1000 / 22 MeV	2
^{8}Li	^{9}Be(^{7}Li,^{8}Li)^{8}Be	+0.367	10^5 -10^6	500 / 25.8 MeV	44
^{8}B	^{3}He(^{6}Li,^{8}B)n	-1.975	10^4	1000 / 15.6 MeV	4.4
^{10}Be	^{9}Be(^{11}B,^{10}Be)^{10}B	-4.642	10^5	800 / 23.2 MeV	3
^{12}B	^{9}Be(^{11}B,^{12}B)^{8}Be	+1.705	10^5	800 / 25.0 MeV	

The solenoids make a selection by the magnetic rigidity of the particles

$$B\rho = \sqrt{2mE}\Big/ q$$

, where m is the mass, E the energy and q the charge state of the particle. The secondary beam focused in the intermediate scattering chamber between the two solenoids (chamber 2), is usually a cocktail of particles, all with same magnetic rigidity but different masses, charge states and energies. These contaminations come mainly from the degraded primary beam, scattered in the primary target, but also from light particles such as alphas, deuterons and protons which are produced by reactions in the primary target. These unwanted particles can be partially suppressed by using blockers and collimators strategically positioned along the solenoid beam line, nevertheless the purity of the secondary beam is usually rather poor in chamber 2. The purity of the secondary beams can be improved by the use a degrador foil placed in the intermediate scattering chamber (chamber 2) and a subsequent selection by the second solenoid. More recently, secondary beams of ^{6}He and ^{8}Li with purity better than *95%* have been obtained in RIBRAS using the double solenoid system [27] and a degrador in chamber 2.

2.2. The ^{6}He + ^{58}Ni Experiment

The ^{6}He + ^{58}Ni experiment has been performed in the intermediate scattering chamber 2, using only 1 solenoid to select the ^{6}He secondary beam. The I ~ 200 nAe ^{7}Li primary beam of energies 18, 22 and 27 MeV was used to produce the ^{6}He secondary beam *via* the ^{9}Be(^{7}Li, ^{6}He)^{10}B production reaction. A 12 μm ^{9}Be foil was used as primary target. Four E(1000 μm)-ΔE(20 μm) telescopes formed by silicon detectors have been mounted in the rotating plate installed in the intermediate scattering chamber. A 2.2 mg/cm^2 ^{58}Ni secondary target was used in the experiment. In Fig. (**2**) we show a bidimensional E-ΔE spectrum where we can see the ^{6}He elastic scattering peak, the contaminations ^{7}Li, α and other lighter particles. We see that the contaminations are well separated from the elastic scattering peak.

For normalization purposes, Gold target runs were performed just before and after every run with the ^{58}Ni target. Since the ^{6}He + ^{197}Au is pure Rutherford at the energies and angles of this experiment the differential cross sections can be obtained from the expression below:

$$\sigma^{^6He+^{58}Ni}\;\theta\;=\frac{N_c^{Ni}}{N_c^{Au}}\frac{N_b^{Au}}{N_b^{Ni}}\frac{N_t^{Au}}{N_t^{Ni}}\frac{J^{Ni}}{J^{Au}}\sigma_{Ruth}^{^6He+^{187}Au}(\theta) \tag{1}$$

Where in equation (1) stands, N_c is the elastic peak integral ^{58}Ni or ^{197}Au, N_b is the number of incident ^{6}He particles during the run, J is the Jacobian to transform from laboratory to center-of-mass frames, and N_t is the density of the secondary target in atoms/cm^2. In this normalization method, we suppose that the secondary beam is proportional to the primary beam intensities, which are measured by the

Faraday cup. This kind of normalization has the advantage of not depending on the solid angles of the detectors. The resulting ^{6}He + ^{58}Ni angular distributions are shown in Fig. (**3**). The experimental errors are due to the statistics in the elastic scattering peak.

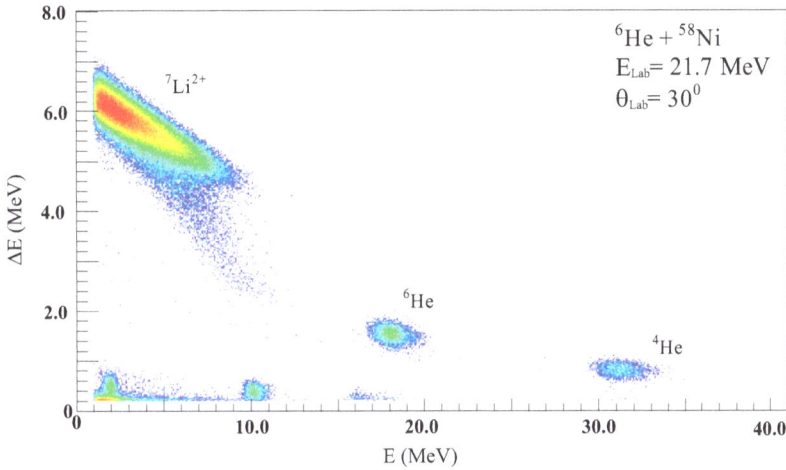

Fig. (2). Two-dimensional spectrum obtained with the ^{6}He beam and ^{58}Ni target at θ_{lab} = 30° and and E_{lab} =21.7 MeV [23].

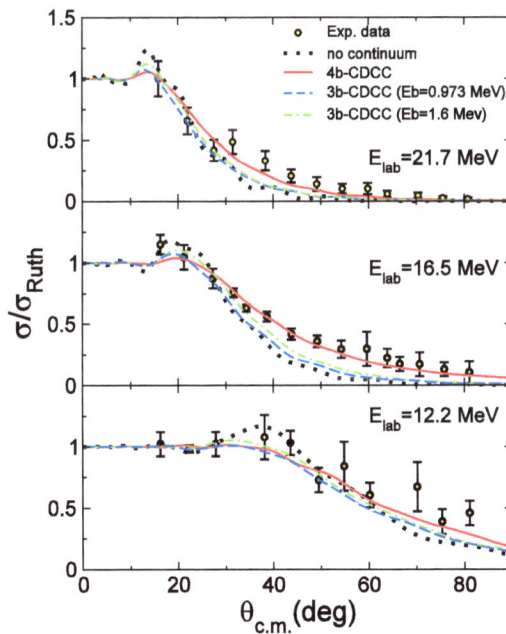

Fig. (3). ^{6}He + ^{58}Ni angular distributions compared with, no-coupling, 4- and 3-body CDCC calculations [23].

THEORETICAL DESCRIPTION

The scattering of a weakly bound nuclei on a (structureless) target, as already mentioned in the introduction, can be analyzed theoretically within the Continuum-Discretized Coupled-Channels (CDCC) formalism [13, 14].

Let us first summarize the basics of the formalism. Considering a N-body system that collides with a structureless target, the total scattering wavefunction is solution of the Schrödinger equation

$$\left[\hat{H}-E\right]\Psi(\mathbf{R},\xi)=0 \tag{2}$$

where **R** is the coordinate from the target to the center of mass of the projectile (as shown in Fig. **4**), ξ are the projectile internal coordinates and E is the center-o--mass scattering energy. The Hamiltonian has the form

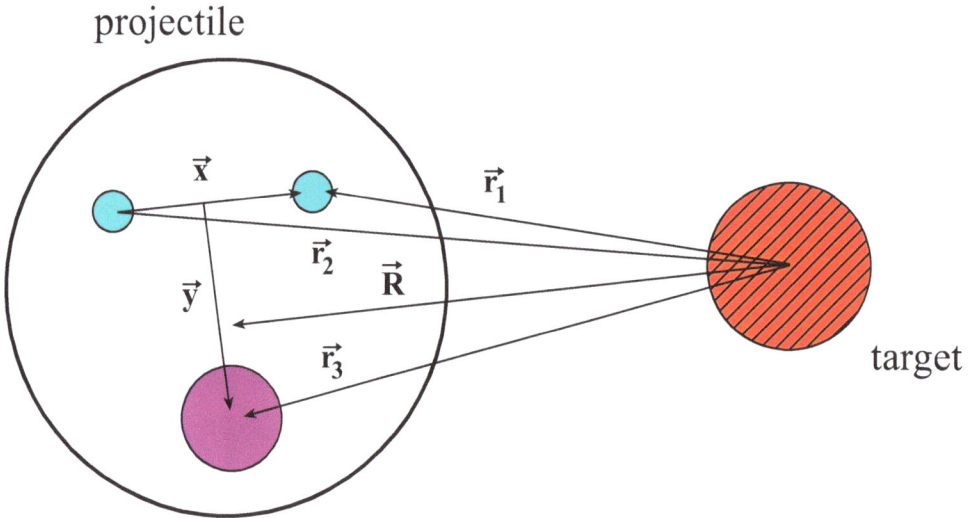

Fig. (4). Relevant coordinates for the scattering of a three-body projectile by a structureless target.

$$\hat{H}=\hat{T}_R+\hat{U}_{pt}+\hat{h} \tag{3}$$

being $\hat{T}_R$ the kinetic energy of the projectile-target relative motion, $\hat{h}$ the internal Hamiltonian of the projectile and $\hat{U}_{pt}$ the interaction between the projectile and the target. The last is assumed to be a sum over the interaction of each projectile-particle with the target

$$\hat{U}_{pt} = \sum_{i=1}^{N} V_{it}(\mathbf{R}_i) \qquad (4)$$

The basic idea of the CDCC formalism consists in expanding the wave-function of the whole system, that is, of the projectile-target system in the states of the projectile. For a total angular momentum J (and projection M) the wave function will be

$$\Psi_{JM}(\mathbf{R},\xi) = \sum_{nj\mu LM_L} \phi_{nj\mu}(\xi) \langle LM_L j\mu \| JM \rangle \frac{i^L}{R} Y_{LM_L}(\hat{R}) f^J_{Lnj}(R) \qquad (5)$$

where $\phi_{nj\mu}$ are the states of the projectile for an angular momentum j (and projection μ). The label n is introduced to enumerate the projectile states and its meaning will depend on the discretization method used to treat the continuum. The projectile angular momentum j is coupled to the orbital angular momentum of the projectile-target relative motion L to obtain J, the total angular momentum of the system. The radial functions $f^J_{Lnj}(R)$ are then obtained introducing the expansion (5) in the Schrödinger equation, leading to a system of coupled equations

$$\left[-\frac{\eta^2}{2m_r}\left(\frac{d^2}{dR^2} - \frac{L(L+1)}{R^2}\right) + E_{nj} - E \right] f^J_{Lnj}(R) + \sum_{L'n'j'} i^{L'-L} V^J_{Lnj,L'n'j'}(R) f^J_{Lnj}(R) = 0 \qquad (6)$$

where m_r is the reduced mass of the projectile-target system and E_{nj} are the energies of the projectile states over the ground state. If ε_{gs} is the energy of the ground with respect to the breakup threshold, the energies of the projectile over the breakup threshold are defined $\varepsilon_{nj} = E_{nj} + \varepsilon_{gs}$. The coupling potentials $V^J_{Lnj,L'n'j'}(R)$ are defined as

$$V^J_{Lnj,L'n'j'}(R) = \langle LnjJM | \hat{V}_{pt} | L'n'j'JM \rangle \qquad (7)$$

where the ket $|LnjJM\rangle$ denotes the function $\Phi_{Lnj}^{JM}(\hat{R},\xi)$ given by

$$\Phi_{Lnj}^{JM}(\hat{R},\xi)=\sum_{\mu M_L}\phi_{nj\mu}(\xi)\langle LM_L j\mu|JM\rangle Y_{LM_L}(\hat{R}) \tag{8}$$

To calculate these coupling potentials, a multipole Q expansion of the projectile-target interaction is usually performed. But, first, it is necessary to obtain the states of the projectile, both the bound states and a discrete representation of the continuum. Note that all the information about the projectile structure is included in the coupling potentials Eq. (7) and the complexity of the matrix elements is getting increased with the number of projectile particles N.

To describe a three-body projectile, such as ^{6}He$(\alpha+n+n)$, we need the Jacobi coordinates $\{\mathbf{x},\mathbf{y}\}$, shown in Fig. (4) . The variable $\mathbf{x}$ is proportional to the relative coordinate between two of the particles and $\mathbf{y}$ is proportional to the coordinate from the center of mass of these two particles to the third particle. From the Jacobi coordinates, the hyperspherical coordinates $\{\rho, \alpha, \}$, can be defined. The hyperradius (ρ) and the hyperangle (α) are given by $\rho=\sqrt{x^2+y^2}$ and $tan\ \alpha = x/y$. Working with the hyperspherical coordinates, we can use the Hyperspherical Harmonics [28] method, i. e., the wave functions of the system are expanded in states with good total angular momentum j:

$$Y_{\beta j\mu}(\Omega)=\left[\left[Y_{Kl}^{l_x l_y}(\Omega)\otimes\chi_{S_x}\right]\otimes\chi_I\right]_{j\mu} \tag{9}$$

where $Y_{Kl}^{l_x l_y}(\Omega)$ are the Hyperspherical Harmonics that depend on the angular variables, $\Omega\equiv\{\alpha,\hat{x},\hat{y}\}$, $\chi_{S_x}^\sigma$ and χ_I^I are the spin functions of the two particles related by the coordinate $\mathbf{x}$ and of the third particle, respectively. Each component of the wavefunction (or channel) is defined by the set of quantum numbers $\beta\equiv\{K,l_x,l_y,l,S_x,j_{ab}\}$. Here, K is the hypermomentum, l_x and l_y are the orbital angular momenta associated with the Jacobi coordinates $\mathbf{x}$ and $\mathbf{y}$, $\mathbf{l}=\mathbf{l}_x+\mathbf{l}_y$ is the total orbital angular momentum, S_x is the spin of the particles related by the coordinate $\mathbf{x}$, and $\mathbf{j}_{ab}=\mathbf{l}+\mathbf{S}_x$. Finally, $\mathbf{j}=\mathbf{j}_{ab}+\mathbf{I}$ is the total angular momentum, with I the spin of the third particle, which we assume fixed. The states of the system in a discrete representation can now be expressed as a linear combination of the states given by Eq. (9) as

$$\phi_{nj\mu}(\mathbf{x},\mathbf{y})=\sum_{\beta}R_{n\beta j}(\rho)Y_{\beta j\mu}(\Omega)\tag{10}$$

where $R_{n\beta j}(\rho)$ are the hyperradial functions. The label n enumerates the state and its meaning will depend on the discretization method.

The binning procedure, traditionally used in three-body CDCC calculations to discretize the continuum of two-body projectiles, consists in calculating first the true continuum wave functions and then making packages or bins in energy or momentum. The extension to three-body projectiles was proposed in Ref [15] within the Hyperspherical Harmonic method. If the three-body continuum wave functions for a momentum κ, related to the continuum energy as $\kappa=\sqrt{2m|\varepsilon|}\big/\eta$ with m a normalization mass, and with incoming conditions given by $\beta`$ are

$$\phi_{\kappa\beta`j\mu}(\mathbf{x},\mathbf{y})=\sum_{\beta}R_{\beta`\beta j}(\kappa\rho)Y_{\beta j\mu}(\Omega)\tag{11}$$

the hyperradial functions for an energy interval or bin is calculated as follows

$$R_{n\beta j}^{bin}(\rho)\equiv R_{[\kappa_1,\kappa_2]\beta`\beta j}^{bin}(\rho)=\frac{2}{\sqrt{\pi N_{\beta`j}}}\int_{\kappa_1}^{\kappa_2}d\kappa\,\omega_{\beta`j}(\kappa)R_{\beta`\beta j}(\kappa\rho)\tag{12}$$

where $R_{\beta`\beta j}(\kappa\rho)$ are the continuum hyperradial wave functions and $\omega_{\beta`j}(\kappa)$ is a weight function with $N_{\beta`j}$ its normalization constant.

Note that in a three-body case we deal with a multi-channel problem, that is, the total wavefunction will be a linear combination of incoming channels $\beta`$ and outgoing channels $\beta$. For this reason, the quantum number $n$ includes here reference to the energy interval of the bin $[\kappa_1, \kappa_2]$, as well as to the incoming channel $\beta`$. It follows that to include a large number of $\beta`$ channels is a severe computational challenge, and that it is desirable to establish a hierarchy of the continuum states according to their importance to the reaction dynamics. In so doing, we may be able to describe scattering observables using only a selected set of states, the number of them depending on the reaction under study. To this end, we make use of the eigenstates of the multi-channel three-body S-matrix, or eigenchannels (EC), as follows. (i) For each angular momentum j and continuum energy ε, the S-matrix in the β basis is diagonalized to obtain its EC, their corresponding eigenvalues and eigenphases. (ii) The magnitudes of these eigenphases are used to order the EC. In Ref [15] we showed that those EC with largest phase shifts are the most strongly coupled in the reaction dynamics, and thus a hierarchy of states can be established by such an ordering. This leads to the possibility of a truncation in the number of the EC included, and testing the convergence with respect to this number of EC cut-off parameter.

APPLICATION TO ^{6}HE + ^{58}NI SCATTERING

The nucleus of ^{6}He is treated as a three-body system comprising an inert α particle core and two valence neutrons. A notable property of ^{6}He is that none of its binary sub-systems bind, while the three-body system has a single bound state with binding energy of 0.973 MeV and total angular momentum $j^\pi = 0^+$. Its low-lying continuum spectrum is dominated by a narrow $j^\pi = 2^+$ resonance, 0.825 MeV above threshold.

The three-body Hamiltonian describing the $\alpha + n + n$ system, includes two-body interactions plus an effective three-body potential. The n- ^{4}He potential was taken from [29] and the NN potential from Ref [30]. The three-body potential is a simple phenomenological three-body force, depending only on the hyperradius, introduced to correct the under-binding caused by the neglect of other configurations. This potential includes parameters which can be fixed to adjust the position of the known states of the system to the experimental values without distorting their structure.

The ground state and continuum wave functions were obtained by solving the Schrödinger equation in hyperspherical coordinates, following the procedure described in Ref [28], and making use of the codes FACE [31] and STURMXX [32]. The maximum hypermomentum was established to $K_{max} = 8$, which provides converged results for the elastic scattering of ^{6}He from a heavy target [19].

The calculated ^{6}He ground state has binding energy of 0.953 MeV and a single-particle density with root mean squared (rms) radius 2.46 fm, assuming an α particle rms radius of 1.47 fm. The continuum states were calculated for angular momenta j= 0^+, 1^-, 2^+, 3^-. As an example, Fig. (**5**) shows the phase shift as a function of the continuum energy for j = 2^+. We include in the figure, the calculation for the first four EC (largest phase shifts). The phase shift corresponding to the first EC presents a characteristic resonant behaviour around 0.85 MeV.

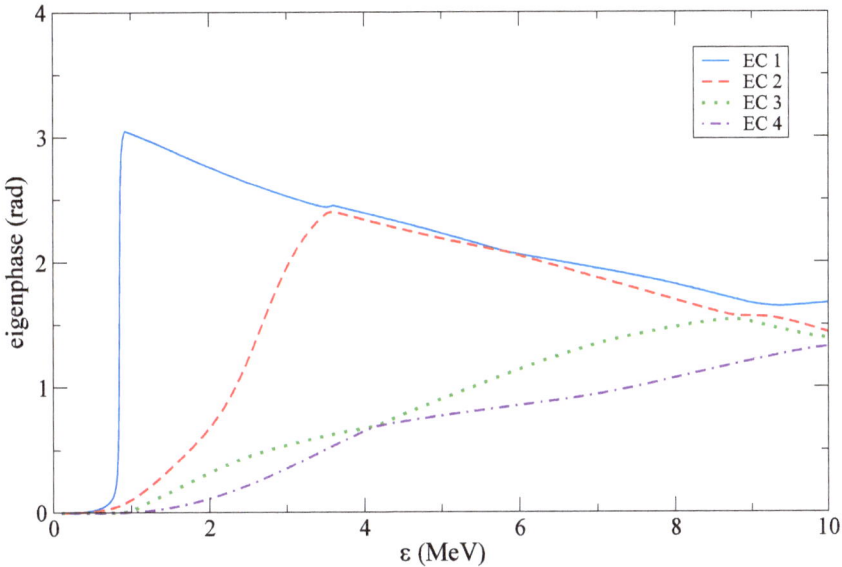

Fig. (5). Phase shift for the first four $j^\pi = 2^-$ eigenchannels (EC) of ^{6}He.

Next, we performed the four-body CDCC calculations for the reaction of ^{6}He on ^{58}Ni at the incident energies 12.2, 16.5, and 21.7 MeV. The optical potentials between the projectile fragments and the target, $\alpha + {}^{58}$Ni and n + ^{58}Ni are shown in Table **2**, where E_{lab} is the energy in Lab frame. V and W are the real and imaginary depths of the optical potential. The reduced radius and diffuseness are r and a, respectively. For the incident energy 12.2 MeV the optical potentials are taken from Refs [33, 34] for $\alpha + {}^{58}$Ni and n + ^{58}Ni, respectively, and for 16.5 and 21.7 MeV are both taken from Ref [35]. We included in the CDCC calculations the states with angular moment j = 0^+, 1^-, 2^+, 3^- up to 7, 8, and 9 MeV for the incident energies 12.2, 16.5, and 21.7 MeV, respectively, in order to get convergence. The maximum EC included was the 4th. The coupling potentials were calculated up to multipole order Q = 3. The coupled equations were solved using the code FRESCO [36], reading externally the coupling potentials. The convergence of the coupled-channels calculations is checked by two criteria: the total cross section as a function of the total angular momentum J must converge (be very small) for high J and the elastic angular distributions are not sensitive to variations in the number of bins, the maximum energy, the number of EC, the multipole order Q and the projectile angular momenta j included in the calculation.

The results of the calculations are shown in Fig. (**3**). The dotted black line is the four-body CDCC without coupling to the continuum. The solid red line corresponds to the full four-body CDCC calculation. We can see from Fig. (3) that the most relevant effect of the coupling to the breakup channel is the flux removal from the forward angles region, reducing the elastic cross section in the region of the Fresnel peak. There is also an increase in the backward angles that improves the agreement with the experimental data.

Table 2. Optical potentials used in the CDCC calculations at the three energies, respectively, for α + ^{58}Ni (top row) and n + ^{58}Ni (bottom row).

E_{lab} (MeV)	V_o (MeV)	r_o (fm)	a_o (fm)	W_o (MeV)	r_i (fm)	a_i (fm)	W_d (MeV)	r_d (fm)	a_d (fm)
12.2	60.00	1.62	0.54	0.50	1.62	0.54	15.84	1.52	0.44
	61.36	1.45	0.57	-	-	-	1.28	1.45	0.50
16.5	165.90	1.62	0.40	11.40	1.62	0.40	23.98	1.52	0.44
	42.00	1.46	0.35	6.09	1.46	0.35	-	-	-
21.7	135.10	1.35	0.64	7.64	1.34	0.50	18.97	1.52	0.44
	42.00	1.46	0.35	6.09	1.46	0.35	-	-	-

To compare with the four-body CDCC calculations, and with the experimental data, we also include in Fig. (3) two three-body CDCC calculations that use a dineutron model for ^{6}He. One uses the standard dineutron model, taking for ^{6}He the experimental binding energy 0.973 MeV, and the second uses the improved dineutron model, taking an effective two-neutron separation energy of 1.6 MeV [24]. For details of these calculations see Ref [23] and references there in. In Fig. (3), the dashed blue line and the dash-dotted green line are the full three-body CDCC calculations for the standard dineutron model and the improved dineutron model, respectively. We can see that, meanwhile the behaviour at forward angles is also given by the three-body CDCC calculations, both three-body calculations fail to reproduce the backward angles rise seen in the data. The improved three-body model becomes closer to the data than the standard three-body, but it is not sufficient to reproduce them.

In Fig. (6) we present, for the incident energy E_{lab} = 16.5 MeV, the local equivalent polarization potential extracted from the CDCC calculations [37] as a function of the distance R between ^{6}He and ^{58}Ni. The polarization from four-body CDCC is shown by solid red line and from standard three-body by a dashed blue line. The real part of the polarization potential is mainly due to the nuclear breakup and it is repulsive in the region of the surface of the system $R \sim$ 6-7 fm . The imaginary part has a long-range absorptive component due to the long-range Coulomb breakup [38].

ANALYSIS OF THE TOTAL REACTION CROSS SECTIONS

The theoretical analysis of the elastic scattering angular distributions provided the total reaction cross sections of ^{6}He + ^{58}Ni at each energy. The total reaction cross section comprises all non-elastic processes such as fusion, inelastic excitations of the projectile and target and transfer reactions which are the most common processes occurring in the collision between stable nuclei at the energies of the present experiment. Complete fusion is in general the most important process in terms of cross section, exhausting most of the total reaction cross section at energies near and a little above the Coulomb barrier. The remaining processes contributing to the total reaction cross section are due to reactions such as inelastic scattering or direct transfer reactions.

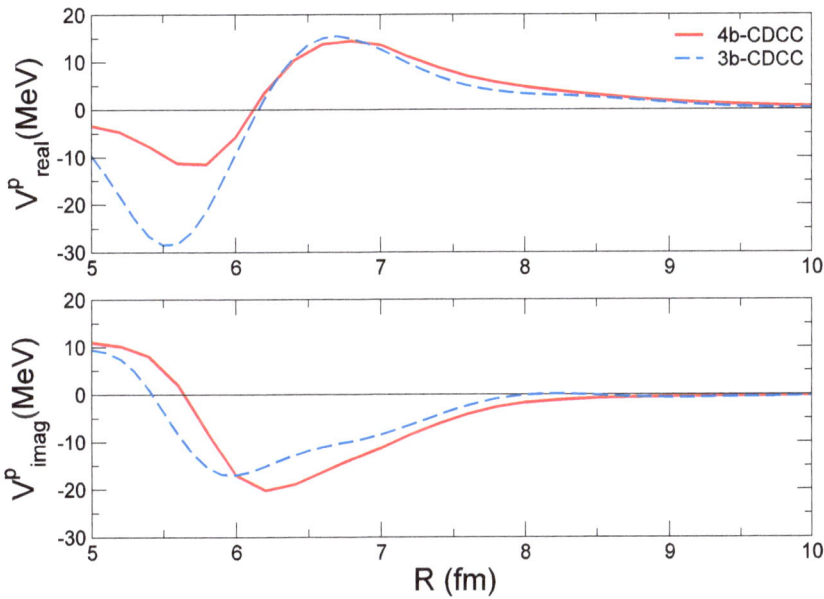

Fig. (6). Polarization potential from four-body and standard three-body CDCC calculations for the incident energy E_{lab} =16.5 MeV.

On the other hand, in the case of collisions involving weakly bound or neutron rich exotic projectiles, other processes can occur with high probability. The projectile breakup reactions, incomplete fusion, neutron transfer reactions are examples. In the incomplete fusion process, a part of the projectile fuses with the target forming a highly excited nucleus which will subsequently decay by emitting particles and gammas, leaving a residual nuclei. In the case of neutron rich halo projectiles, such as ^{6}He, neutron transfer reactions from the projectile to the target can be induced even at energies below the barrier since the neutrons do not feel the repulsive Coulomb barrier. The strong dipolar polarizability of ^{6}He may also induce deformations in the projectile along the incoming trajectories, facilitating neutron transfer and projectile breakup reactions. In some cases the neutrons emitted by the projectile breakup can be subsequently absorbed by the target, forming a highly excited system. In this picture, there is no clear distinction between processes such as incomplete fusion, neutron transfer and projectile breakup. Kinematical and Q-optimum arguments can be applied to try to distinguish between the processes but, in most cases, what is observed is an intermediate situation involving different contributions to the total reaction cross section of exotic systems [39, 40]. As a result, the total reaction cross sections are expected to be larger for exotic systems than for stable ones.

In Fig. (7) we present total reaction cross sections for several stable, weakly bound and exotic systems of intermediate masses around A = 60. We plot the reduced cross sections and energies which are dimensionless quantities defined as: $\sigma_{red} = \sigma/\pi R_b^2$ and $E_{red} = E_{c.m.}/V_b$ where R_b and V_b are respectively the Coulomb barrier radius and strength calculated by the formulas developed in reference [41]. This kind of reduction method aims to remove trivial effects such as different radius and different Coulomb barriers in order to compare systems of different masses all in the same plot.

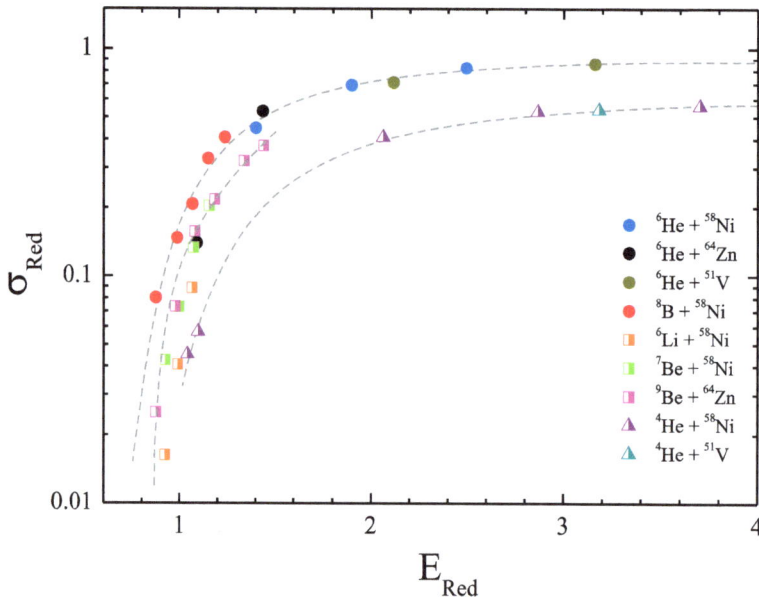

Fig. (7). Reduced reaction cross sections as a function of the reduced energies for tightly bound, weakly bound and exotic system with masses A ~ 60.

The results show that there are three classes of reduced cross sections, increasing from the tightly bound to the stable weakly bound and finally the exotic systems with ^{6}He projectile which presents the highest reaction cross sections. This enhancement effect in the total reaction cross sections as the projectile becomes more and more exotic has been observed previously in heavier and lighter systems [12, 38, 42].

CONCLUDING REMARKS AND FUTURE PROSPECTS

^{6}He + ^{58}Ni elastic scattering angular distributions have been measured at three energies slightly above the Coulomb barrier. The data have been analyzed by the *state of the art* four- and three-body Continuum - Discretized Coupled Channels-Calculations (CDCC) where the coupling between the elastic scattering and the breakup of the projectile is taken into account explicitly.

The four-body CDCC results, which takes into account the full three-body Borromean structure of the ^{6}He projectile, reproduced the data very well with no free parameters. However, an enhancement in the differential cross sections observed at intermediate angles, is not reproduced by the three-body CDCC calculations, that uses a dineutron model for ^{6}He.

The total reaction cross sections for ^{6}He + ^{58}Ni have been obtained from the CDCC calculations, and are compared in a systematics involving other targets of similar masses with tightly bound and weakly bound projectiles. In order to remove trivial effects due to different masses and Coulomb barriers, a reduction method was applied to the cross sections and energies. A single plot of all reduced cross sections and energies show that there are three classes of reduced cross sections, increasing from tightly bound to weakly bound an finally for the exotic ^{6}He, which presents the highest cross section. The enhancement observed in the total reaction cross section of ^{6}He is probably a consequence of the exotic nature of the projectile and was observed previously in the ^{6}He scattering on heavier targets. RIB research has been carried out so far in first generation facilities where the intensities of the secondary beams were limited to 10^{5-7} pps, several orders of magnitude below the intensities of the present primary beams. As a consequence, the experimental data have large error bars when compared to the errors of the data along the stability line. Nevertheless, many interesting results have been obtained about the nature of these exotic species.

On the other hand, new theoretical techniques have been developed along with the experimental achievements which allow to perform coupled channels calculations taking into account the effect of the continuum of the weakly bound projectiles. The remarkable enhancement in the calculation capabilities of the new computers allows to perform coupled channels calculations involving a number of states that was unthinkable decades ago.

New generation RIB facilities are already running or are expected to enter into operation in the next years which promise an increase of several orders of magnitude in the intensities of the secondary beams. This will represent a qualitative leap in the accuracy of experimental data from forthcoming experiments. The developments in RIB research in the next years seem to be quite promising from both, experimental and theoretical points of view. In spite of the fast evolving research capabilities in large national laboratories all over the World, small facilities such as RIBRAS and others can play a role in the overall picture, improving their detecting capabilities and performing their research in energy domains which are sometimes overlooked by the large laboratories.

CONSENT FOR PUBLICATION

Not applicable.

CONFLICT OF INTEREST

The author (editor) declares no conflict of interest, financial or otherwise.

ACKNOWLEDGMENTS

The authors acknowledge the Brazilian funding agencies *Fundação de Amparo a Pesquisa no Estado de São Paulo (FAPESP)* grant 2013/22100-7, and *Conselho Nacional de Pesquisa (CNPq)* for the financial support. This work has been partially supported by the Spanish Ministerio de Economía y Competitividad and the European Regional Development Fund (FEDER) under Project No. FIS2014-51941-P. M. Rodríguez-Gallardo acknowledges postdoctoral support from the Universidad de Sevilla under the V Plan Propio de Investigacíon contract No. USE-11206-M.

REFERENCES

[1]   H. Simon, D. Aleksandrov, T. Aumann, L. Axelsson, T. Baumann, M.J.G. Borge, L.V. Chulkov, R. Collatz, J. Cub, W. Dostal, B. Eberlein, Th.W. Elze, H. Emling, H. Geissel, A. Grünschloss, M. Hellström, J. Holeczek, R. Holzmann, B. Jonson, J.V. Kratz, G. Kraus, R. Kulessa, Y. Leifels, A. Leistenschneider, T. Leth, I. Mukha, G. Munzenberg, F. Nickel, T. Nilsson, G. Nyman, B. Petersen, M. Pfutzner, A. Richter, K. Riisager, C. Scheidenberger, W. Schwab, M.H. Smedberg, J. Stroth, A. Surowiec, O. Tengblad, and M.V. Zhukov, "Direct experimental evidence for strong admixture of different parit states 11Li", *Phys. Rev. Lett.,* vol. 83, pp. 496-499, 1999. [http://dx.doi.org/10.1103/PhysRevLett.83.496]

[2]   T. Otsuka, T. Suzuki, R. Fujimoto, H. Grawe, and Y. Akaishi, "Evolution of nuclear shells due to the tensor force", *Phys. Rev. Lett.,* vol. 95, no. 23, p. 232502, 2005. [http://dx.doi.org/10.1103/PhysRevLett.95.232502] [PMID: 16384301]

[3]   A. Bartlett, J. Görres, J.G. Mathews, K. Otsuki, M. Wiescher, D. Frekers, A. Mengoni, and J. Tostevin, "Two-neutron capture reactions and the r process", *Phys. Rev. C Nucl. Phys.,* vol. 74, p. 015802, 2006. [http://dx.doi.org/10.1103/PhysRevC.74.015802]

[4]   I. Tanihata, H. Hamagaki, O. Hashimoto, Y. Shida, N. Yoshikawa, K. Sugimoto, O. Yamakawa, T. Kobayashi, and N. Takahashi, "Measurements of interaction cross sections and nuclear radii in the light p-shell region", *Phys. Rev. Lett.,* vol. 55, no. 24, pp. 2676-2679, 1985.
[http://dx.doi.org/10.1103/PhysRevLett.55.2676] [PMID: 10032209]

[5]   V. Efimov, "Energy levels arising from resonant two-body forces in a three-body system", *Phys. Lett. B,* vol. 33, pp. 563-564, 1970.
[http://dx.doi.org/10.1016/0370-2693(70)90349-7]

[6]   T. Kraemer, M. Mark, P. Waldburger, J.G. Danzl, C. Chin, B. Engeser, A.D. Lange, K. Pilch, A. Jaakkola, H.C. Nägerl, and R. Grimm, "Evidence for Efimov quantum states in an ultracold gas of caesium atoms", *Nature,* vol. 440, no. 7082, pp. 315-318, 2006.
[http://dx.doi.org/10.1038/nature04626] [PMID: 16541068]

[7]   M. Cubero, J.P. Fernández-García, M. Rodríguez-Gallardo, L. Acosta, M. Alcorta, M.A.G. Alvarez, M.J.G. Borge, L. Buchmann, C.A. Diget, H. Al Falou, B.R. Fulton, H.O.U. Fynbo, D. Galaviz, J. Gómez-Camacho, R. Kanungo, J.A. Lay, M. Madurga, I. Martel, A.M. Moro, I. Mukha, T. Nilsson, A.M. Sánchez-Benítez, A. Shotter, O. Tengblad, and P. Walden, "Do halo nuclei follow Rutherford elastic scattering at energies below the barrier? The case of 11Li", *Phys. Rev. Lett.,* vol. 109, no. 26, p. 262701, 2012.
[http://dx.doi.org/10.1103/PhysRevLett.109.262701] [PMID: 23368554]

[8]   A. Di Pietro, G. Randisi, V. Scuderi, L. Acosta, F. Amorini, M.J.G. Borge, P. Figuera, M. Fisichella, L.M. Fraile, J. Gomez-Camacho, H. Jeppesen, M. Lattuada, I. Martel, M. Milin, A. Musumarra, M. Papa, M.G. Pellegriti, F. Perez-Bernal, R. Raabe, F. Rizzo, D. Santonocito, G. Scalia, O. Tengblad, D. Torresi, A. Maira Vidal, D. Voulot, F. Wenander, and M. Zadro, "Elastic scattering and reaction mechanisms of the halo nucleus 11Be around the Coulomb barrier", *Phys. Rev. Lett.,* vol. 105, no. 2, p. 022701, 2010.
[http://dx.doi.org/10.1103/PhysRevLett.105.022701] [PMID: 20867705]

[9]   E.F. Aguilera, J.J. Kolata, F.M. Nunes, F.D. Becchetti, P.A. DeYoung, M. Goupell, V. Guimarães, B. Hughey, M.Y. Lee, D. Lizcano, E. Martinez-Quiroz, A. Nowlin, T.W. O'Donnell, G.F. Peaslee, D. Peterson, P. Santi, and R. White-Stevens, "Transfer and/or breakup modes in the 6He + 209Bi reaction near the coulomb barrier", *Phys. Rev. Lett.,* vol. 84, no. 22, pp. 5058-5061, 2000.

[10] E.F. Aguilera, J.J. Kolata, F.D. Becchetti, P.A. DeYoung, J.D. Hinnefeld, Á. Horváth, L.O. Lamm, H.Y. Lee, D. Lizcano, E. Martinez-Quiroz, P. Mohr, T.W. O'Donnell, D.A. Roberts, and G. Rogachev, "Elastic scattering and transfer in the 6He + 209Bi system below the Coulomb barrier", *Phys. Rev. C Nucl. Phys.,* vol. 63, p. 061603, 2001. [R].
[http://dx.doi.org/10.1103/PhysRevC.63.061603]

[11] L. Acosta, A.M. Sánchez-Benítez, M.E. Gómez, I. Martel, F. Pérez-Bernal, F. Pizarro, J. Rodríguez-Quintero, K. Rusek, M.A.G. Alvarez, M.V. Andrés, J.M. Espino, J.P. Fernández-García, J. Gómez-Camacho, A.M. Moro, C. Angulo, J. Cabrera, E. Casarejos, P. Demaret, M.J.G. Borge, D. Escrig, O. Tengblad, S. Cherubini, P. Figuera, M. Gulino, M. Freer, C. Metelko, V. Ziman, R. Raabe, I. Mukha, D. Smirnov, O.R. Kakuee, and J. Rahighi, "Elastic scattering and α-particle production in $ 6He + 208Pb collisions at 22 MeV", *Phys. Rev. C Nucl. Phys.,* vol. 84, p. 044604, 2011.
[http://dx.doi.org/10.1103/PhysRevC.84.044604]

[12] P.N. de Faria, R. Lichtenthäler, K.C.C. Pires, A.M. Moro, A. Lépine-Szily, V. Guimarães, D.R. Mendes, A. Arazi, M. Rodríguez-Gallardo, A. Barioni, V. Morcelle, M.C. Morais, O. Camargo, J. Alcantara Nuñez, and M. Assunção, "Elastic scattering and total reaction cross section of 6He + 120Sn", *Phys. Rev. C Nucl. Phys.,* vol. 81, p. 044605, 2010.
[http://dx.doi.org/10.1103/PhysRevC.81.044605]

[13] M. Yahiro, Y. Iseri, H. Kameyama, M. Kamimura, and M. Kawai, "Effects of deuteron virtual breakup on deuteron elastic and inelastic scattering", *Prog. Theor. Phys. Suppl.,* vol. 89, pp. 32-86, 1986.
[http://dx.doi.org/10.1143/PTPS.89.32]

[14] N. Austern, Y. Iseri, M. Kamimura, M. Kawai, G. Rawitscher, and M. Yahiro, "Continuum-discretized

coupled-channels calculations for three-body models of deuteron-nucleus reactions", *Phys. Rep.,* vol. 154, pp. 125-204, 1987.
[http://dx.doi.org/10.1016/0370-1573(87)90094-9]

[15]   M. Rodríguez-Gallardo, J.M. Arias, J. Gómez-Camacho, A.M. Moro, I.J. Thompson, and J.A. Tostevin, "Four-body continuum-discretized coupled-channels calculations", *Phys. Rev. C Nucl. Phys.,* vol. 80, p. 051601, 2009. [R].
[http://dx.doi.org/10.1103/PhysRevC.80.051601]

[16]   T. Matsumoto, T. Kamizato, K. Ogata, Y. Iseri, E. Hiyama, M. Kamimura, and M. Yahiro, "New treatment of breakup continuum in the method of continuum discretized coupled channels", *Phys. Rev. C Nucl. Phys.,* vol. 68, p. 064607, 2003.
[http://dx.doi.org/10.1103/PhysRevC.68.064607]

[17]   A.M. Moro, F. Pérez-Bernal, J.M. Arias, and J. Gómez-Camacho, "Coulomb breakup in a transformed harmonic oscillator basis", *Phys. Rev. C Nucl. Phys.,* vol. 73, p. 044612, 2006.
[http://dx.doi.org/10.1103/PhysRevC.73.044612]

[18]   T. Matsumoto, T. Egami, K. Ogata, Y. Iseri, M. Kamimura, and M. Yahiro, "Coulomb breakup effects on the elastic cross section of 6He + 209Bi scattering near Coulomb barrier energies", *Phys. Rev. C Nucl. Phys.,* vol. 73, p. 051602, 2006. [R].
[http://dx.doi.org/10.1103/PhysRevC.73.051602]

[19]   M. Rodríguez-Gallardo, J.M. Arias, J. Gómez-Camacho, R.C. Johnson, A.M. Moro, I.J. Thompson, and J.A. Tostevin, "Four-body continuum-discretized coupled-channels calculations using a transformed harmonic oscillator basis", *Phys. Rev. C Nucl. Phys.,* vol. 77, p. 064609, 2008.
[http://dx.doi.org/10.1103/PhysRevC.77.064609]

[20]   P. Descouvemont, T. Druet, L.F. Canto, and M.S. Hussein, "Low-energy 9Be + 208Pb scattering, breakup, and fusion within a four-body model", *Phys. Rev. C Nucl. Phys.,* vol. 91, p. 024606, 2015.
[http://dx.doi.org/10.1103/PhysRevC.91.024606]

[21]   J. Casal, M. Rodríguez-Gallardo, and J.M. Arias, "9Be elastic scattering on 208Pb and 27Al within a four-body reaction framework", *Phys. Rev. C Nucl. Phys.,* vol. 92, p. 054611, 2015.
[http://dx.doi.org/10.1103/PhysRevC.92.054611]

[22]   N.C. Summers, F.M. Nunes, and I.J. Thompson, "Extended continuum discretized coupled channels method: Core excitation in the breakup of exotic nuclei", *Phys. Rev. C Nucl. Phys.,* vol. 74, p. 014606, 2006.
[http://dx.doi.org/10.1103/PhysRevC.74.014606]

[23]   V. Morcelle, K.C.C. Pires, M. Rodríguez-Gallardo, R. Lichtenthäler, A. Lépine-Szily, V. Guimarães, P.N. de Faria, D.R. Mendes Junior, A.M. Moro, L.R. Gasques, E. Leistenschneider, R. Pampa Condori, V. Scarduelli, M.C. Morais, A. Barioni, J.C. Zamora, and J.M.B. Shorto, "Four-body effects in the 6He + 58Ni scattering", *Phys. Lett. B,* vol. 732, pp. 228-232, 2014.

[24]   A.M. Moro, K. Rusek, J.M. Arias, J. Gómez-Camacho, and M. Rodríguez-Gallardo, "Improved di-neutron cluster model for 6He scattering", *Phys. Rev. C Nucl. Phys.,* vol. 75, p. 064607, 2007.
[http://dx.doi.org/10.1103/PhysRevC.75.064607]

[25]   R. Lichtenthäler, A. Lépine-Szily, V. Guimarães, C. Perego, V. Placco, O. Camargo Jr, R. Denke, P.N. de Faria, E.A. Benjamim, N. Added, G.F. Lima, M.S. Hussein, J. Kolata, and A. Arazi, "Radioactive Ion beams in Brazil (RIBRAS)", *Eur. Phys. J. A,* vol. 25, pp. 733-736, 2005.
[http://dx.doi.org/10.1140/epjad/i2005-06-043-y]

[26]   A. Lépine-Szily, R. Lichtenthäler, and V. Guimarães, "The Radioactive Ion Beams in Brazil (RIBRAS) facility", *Eur. Phys. J. A,* vol. 50, p. 128, 2014.
[http://dx.doi.org/10.1140/epja/i2014-14128-4]

[27]   R. Lichtenthäler, M.A.G. Alvarez, A. Lépine-Szily, S. Appannababu, K.C.C. Pires, U.U. da Silva, V. Scarduelli, R.P. Condori, and N. Deshmukh, "RIBRAS: The Facility for Exotic Nuclei in Brazil", *Few-Body Syst.,* vol. 57, pp. 157-163, 2016.

[http://dx.doi.org/10.1007/s00601-015-1039-z]

[28]   M.V. Zhukov, B.V. Danilin, D.V. Fedorov, J.M. Bang, I.J. Thompson, and J.S. Vaagen, "Bound state properties of Borromean halo nuclei: 6He and 11Li", *Phys. Rep.,* vol. 231, pp. 151-199, 1993.
[http://dx.doi.org/10.1016/0370-1573(93)90141-Y]

[29]   J. Bang, and C. Gignoux, "A realistic three-body model of 6Li with local interactions", *Nucl. Phys. A.,* vol. 313, pp. 119-140, 1979.
[http://dx.doi.org/10.1016/0375-9474(79)90571-2]

[30]   D. Gogny, P. Pires, and R. de Tourreil, "A smooth realistic local nucleon-nucleon force suitable for nuclear Hartree-Fock calculations", *Phys. Lett. B,* vol. 32, pp. 591-595, 1970.
[http://dx.doi.org/10.1016/0370-2693(70)90552-6]

[31]   I.J. Thompson, F.M. Nunes, and B.V. Danilin, "FaCE: a tool for three body Faddeev calculations with core excitation", *Comput. Phys. Commun.,* vol. 161, pp. 87-107, 2004.
[http://dx.doi.org/10.1016/j.cpc.2004.03.007]

[32]   I.J.   Thompson,   User   Manual,   2004.   Available   in   http://www.fresco.org.uk/programs/sturmxx/index.html.

[33]   M. Avrigeanu, A.C. Obreja, F.L. Roman, V. Avrigeanu, and W. von Oertzen, "Complementary optical-potential analysis of α-particle elastic scattering and induced reactions at low energies", *At. Data Nucl. Data Tables,* vol. 95, pp. 501-532, 2009.
[http://dx.doi.org/10.1016/j.adt.2009.02.001]

[34]   B.A. Watson, P.P. Singh, and R.E. Segel, "Optical-model analysis of nucleon scattering from $1p$-shell nuclei between 10 and 50 MeV", *Phys. Rev.,* vol. 182, pp. 977-989, 1969.
[http://dx.doi.org/10.1103/PhysRev.182.977]

[35]   C.M. Perey, and F.G. Perey, "Compilation of phenomenological optical-model parameters 1954 - 1975", *At. Data Nucl. Data Tables,* vol. 17, pp. 1-101, 1976.
[http://dx.doi.org/10.1016/0092-640X(76)90007-3]

[36]   I.J. Thompson, *Comput. Phys. Commun.,* vol. 7, pp. 167-212, 1988.

[37]   I.J. Thompson, M.A. Nagarajan, J.S. Lilley, and M.J. Smithson, "The threshold anomaly in 16O + 208Pb scattering", *Nucl. Phys. A.,* vol. 505, pp. 84-102, 1989.

[38]   K.C.C. Pires, R. Lichtenthäler, A. Lépine-Szily, V. Guimarães, P.N. de Faria, A. Barioni, D.R. Mendes Junior, V. Morcelle, R. Pampa Condori, M.C. Morais, J.C. Zamora, E. Crema, A.M. Moro, M. Rodríguez-Gallardo, M. Assunção, J.M.B. Shorto, and S. Mukherjee, "Experimental study of 6He + 9Be elastic scattering at low energies", *Phys. Rev. C Nucl. Phys.,* vol. 83, p. 064603, 2011.
[http://dx.doi.org/10.1103/PhysRevC.83.064603]

[39]   P.N. de Faria, R. Lichtenthäler, K.C.C. Pires, A.M. Moro, A. Lépine-Szily, V. Guimarães, D.R. Mendes, A. Arazi, A. Barioni, V. Morcelle, and M.C. Morais, "α-particle production in 6He + 120 Sn collisions", *Phys. Rev. C Nucl. Phys.,* vol. 82, p. 034602, 2010.
[http://dx.doi.org/10.1103/PhysRevC.82.034602]

[40]   K.C.C. Pires, S. Appannababu, R. Lichtenthäler, and O.C.B. Santos., "New method to calculate the nuclear radius from low energy fusion and total reaction cross sections", Phys. Rev. C Nucl. Phys.
[http://dx.doi.org/10.1103/PhysRevC.98.014614]

[41]   A.S. Freitas, L. Marques, X. X. Zhang, M. A. Luzio, P. Guillaumon, R. Pampa Condori, and R. Lichtenthäler, "Woods-Saxon Equivalent to a Double Folding Potential", *Braz J. Phys.,* vol. 46, pp. 120-128, 2016.

[42]   K.C.C. Pires, R. Lichtenthäler, A. Lépine-Szily, and V. Morcelle, "Total reaction cross section for the 6He + 9Be system", *Phys. Rev. C Nucl. Phys.,* vol. 90, p. 027605, 2014.
[http://dx.doi.org/10.1103/PhysRevC.90.027605]

CHAPTER 9

Twin Atoms from Doubly Excited States of the Hydrogen Molecule Induced by Electron Impact

Ginette Jalbert[1,*], **J. Robert**[2], **F. Zappa**[3], **C. R. de Carvalho**[1], **Aline Medina**[4], **L. O. Santos**[4], **F. Impens**[1] and **N.V. de Castro Faria**[1]

[1] *Instituto de Física, UFRJ, Cx. Postal 68528, Rio de Janeiro, RJ21941-972, Brazil*

[2] *Lab. Aimée Cotton CNRS, Univ. Paris Sud-11/ENS Cachan - 91405Orsay, France*

[3] *Departamento de Física, UFJF, MG 36036-330, Brazil*

[4] *Instituto de Física, UFBA, Salvador, BA40210-340, Brazil*

Abstract: Twin photons, pairs of photons with entangled properties, are now easily produced from nonlinear optical crystals and experiments with these pairs are routinely done at different laboratories [1]. Is it possible to study the same kind of properties with twin atoms, *i.e.*, pairs of massive particles also with entangled properties obtained simply by breaking diatomic homonuclear molecules? In this article we discuss experiments, calculations and proposal that point to this possibility.

Keywords: Double excited states, Electron impact, Entanglement, EPR, Hydrogen molecule, Nonlinear optics, Twin atoms, Twin photon.

INTRODUCTION

In 1951 David Bohm [2] provided an alternative and more suitable scheme to accomplish the conditions discussed in the EPR paper [3]. Such experiment would use a pair of atoms, corresponding to a pair of spin-(1/2) particles coming from a molecular fragmentation, in order to probe the existence of non-classical correlations between the particles spins.

Although this "thinking" experiment proposed a pair of atoms, actual EPR pairs were never done with atoms. Indeed, the first experiment by Wu and Shaknov in 1950 was on disintegration of positronium in two photons [4]. Latter, Clauser and Shimony [5], Aspect, Dalibard and Roger [6] and Aspect, Grangier and Roger [7], among others, used pairs of photons. The experiments with fragments of a mole-

---
* **Corresponding author Ginette Jalbert:** Instituto de Física, UFRJ, Cx. Postal 68528, Rio de Janeiro, RJ 21941-972, Brazil; Tel: +55 (21) 3938-7467; Fax: +55 21 3938-7368; E-mail: ginette@if.ufrj.br

Antônio Carlos Fontes dos Santos (Ed.)

cule can be complementary to these approaches, first because of the question of the localization of the various types of particles, second as for a pair of atoms the interaction will browse the complex field of the molecular interactions from the short distances (electrostatics and exchange) to the long distances (Casimir-Polder between moving atoms). The coherence of spin will therefore contain information related to these different interactions and will make use of the dynamic terms related to the movement of the atoms. More precisely, the proposed experiments aim to observe and manipulate the phase of a dissociated molecular state. This state carries out an entangled state naturally, and reveals in a non-trivial way all the dynamic complexity of the system.

Beyond usual tests in this type of EPR experiment, it is interesting to follow, most finely as possible, the dynamics of the dissociation. The molecule that we have chosen is the least complex of the neutral molecules since it is made up only of four bodies. The complexity of symmetries by permutation is nevertheless present since it has two electrons and two protons, but it has the advantage that it can dissociate in a pair of twin atoms. The ionic/covalent configuration also intervenes in dynamics. The couplings of the dissociation channels as twins with these states can, according to the situations, to intervene in the loss of coherence between the twin atoms. Finally, its structure and its dynamics are the subject of renewed studies, as well experimentally by synchrotron radiation or laser techniques, as theoretical by ab initio calculations or by the method of the molecular quantum defect [8 - 10].

In order to analyze the conditions to make the complex experiment proposed, we have done first a systematic study of the fast and the slow $H(2^2S)$ coming from the dissociation of cold hydrogen molecules. It is an experimental benefit that is possible to clearly separate what is called fast and slow fragments. In fact, they were studied from 1967 to now. In the article of M. Leventhal, R. T. Robiscoe and K. R. Lea [11], a time-of-flight metastable $H(2^2S)$ fragments resulting from dissociation of H_2 induced by electrons was employed. They were the first to observe the existence of a group of fast atoms with energies centred at about 5 eV and a group of slow atoms with energies centred at about 0.3 eV. In particular, the group of slow atoms comes in part from the predissociation of molecules in bound vibrational levels of simple excited electronic states that mix at small internuclear distances with dissociating state that asymptotically gives two atoms in ground states and $H(2^2S)$. A different way to form slow $H(2^2S)$ is the direct excitation from the ground state to the same dissociative state energy curves. The group of fast $H(2^2S)$ atoms comes from repulsive doubly excited states known as Q_1 and Q_2 states, that can dissociate in $H(1s) + H(2s)$ and $H(2s) + H(n=2)$.

An important point to be studied is the experimental condition that takes into account the kinematical effects of the collisions. For thermal jet or gas cell, these effects are not visible, but with cold supersonic jet where the average speed of the molecules is about 2.7 km/s, we are in a situation where the electron momentum is of the same order of magnitude of the H_2 momentum. For the slow atom case, the effect on the peak width is the most important aspect to be analysed. For the twin atom experiment with fast metastable atom the kinematics effects do not give origin to peak enlargement, but are crucial in the choice of the detector position.

Apart of our particular interest, the study of neutral fragmentation of H_2 is still of significance. Recently, a theoretical study of the doubly excited states of H_2 converging to the H(n=2) + H(n'=2) limit was performed [12]. The calculations furnish potential energy curves for all internuclear distances. Moreover, the calculations also revealed that a term not included in earlier studies, *i.e.*, a dipole-quadrupole term, is significant at large distances. Another recent article also discusses the dissociation of H_2 [13], but it is essentially interested in symmetry breaking in the case of H_2 excitation by a polarised photon.

The inverse process, the formation of H_2 from collisions of two H(2s) atoms, has recently been the object of study, both theoretically, for atoms at thermal temperatures [14] and experimentally, at low temperatures [15]. Although they give information about the process, we still not know one important value for our future coincidence experiment, the relation between the probabilities to form the channels H(2s) + H(2s) and H(2s) + H(2p). There is little work about the dissociation channel H(2s) + H(2s) [16]. In fact, neutral dissociation is a good probe to investigate molecular doubly excited states. From the applied point of view, the emission from molecular hydrogen is an important diagnostic probe of the peripheral regions of high-temperature fusion plasmas [17].

In fact, it is nevertheless appealing to go back to Bohm's original idea of testing non-classical correlations with the spin observables of massive particles. In fact, quantum concepts applied to separated systems were also discussed in 1935 by Schrödinger [18]. At the beginning of this work he wrote: *"When two systems, of which we know the states by their representatives, enter into temporary physical interaction due to known forces between them, and when after a time of mutual influence, the systems separate again, then they can no longer be described in the same way as before, viz, by endowing each of them with a representative of its own. I would not call that one but rather the characteristic trait of quantum mechanics, the one that enforces its entire departure from classical lines of thought. By the interaction the two representatives (or Ψ–functions) have become entangled".*

The dissociation of a diatomic molecule in a pair of atoms is a well-known problem, and it is also established that the two fragments, named twin atoms for a homonuclear diatomic molecule, will share some coherence between them because they are linked to the same molecular state. The first signatures of the coherence between the fragments is revealed in coincidence experiments, as will be show in this article. The doubly excited states of the H_2 molecules is a good candidate for performing twin atoms experiments because they could dissociate in two H(2s) atomic fragments [19]. The metastable H(2s) is very convenient for studies with Stern-Gerlach atomic spectrometry, for example, having a mean life of 0.1s, compared with another possible fragment, H(2p), that has a nanosecond mean life. In a recent paper [20], the group of the Institut d'Optique (France) was able to obtain twin particles issued from a Bose-Einstein condensate and reproducing the well-known results of Hong-Ou-Mandel interference [21]. We investigate a different way of producing entanglement between massive particles, by going back to Bohm's original proposal involving molecular dissociation.

As this paper introduce concepts of atom optics, we decide to discuss briefly here how atomic fragments propagate in free space. In fact, we give a brief overview of the theoretical tools in atom optics that can be applied to the propagation of the atomic fragments issued from the dissociation of a molecule. These techniques are also applicable to matter-wave interferometers involving entire molecules [22, 23].

When an atomic sample is dilute enough as to neglect the effects of interactions between atoms, the Schrodinger equation takes a form similar to the paraxial equation of an optical field. This assumption is typically met in the experiment considered here, where the fragments are obtained from molecular dissociation at a density which is eight orders of magnitude lower than the atmospheric pressure. This motivates the use of analytical methods borrowed from linear Gaussian optics [24] to describe conveniently the propagation of matter waves.

The ABCD matrices have been adapted from light optics to atom optics by Bordé [25, 26]. One needs to introduce an additional dimension and switch the propagation along the optical axis for a propagation in time. This formalism has been conveniently applied to discuss several experiments, including atom gravimeter [27], atom gyrometer [28], or space-time sensors using the quantum levitation [29 - 31]. The ABCD propagation method shows that Hermite-Gaussian modes have a self-similar propagation in quadratic potentials – which is typically the case for the free-propagation of molecules where only a quasi-uniform gravity field is at work. By linearity of the propagation equation, and by using a decomposition of the initial wave-function on the basis of Hermite-Gauss modes, one can obtain the evolution of an arbitrary atomic wave-packet at any later time.

Another remarkable feature of the ABCD matrices is that these matrices are symplectic. Thus, they preserve the phase-space density of the atomic wave-packet during the propagation. This property enables the definition of an atomic beam quality factor [32, 33] by analogy with optics. The ABCD formalism has been extended to account for atomic samples with moderate interactions [34], or to guided atom lasers [35, 36]. A theory of atom-optical ABCD matrices in a 5-dimensional space has been developed, in which the mass of the atoms becomes itself an observable conjugated to the proper time [37].

Besides the ABCD matrices describing the beam width, one must also the atomic phase acquired by the wave-packet during the propagation. For Gaussian atomic wave-packets evolving in quadratic potentials, this phase is proportional to the action evaluated on the classical trajectory – the trajectory of the center of the atomic wave-packet -. A heuristic explanation of this property is that the path-integral describing the atomic propagation in this case can be reduced to a series Gaussian integrals, and only the phase associated to the stationary classical trajectory survives. For atomic wave-packets evolving in non-quadratic potential, the use of an eikonal approximation can be suitable to estimate the atomic wave-front [38]. Then, one evaluates the action on a series of rays, enabling one to describe properly the atomic wave-front. This technique may be applied in region where the atomic fragments are subject to the influence of the magnetic and electric field with a strong departure from the quadratic potentials. Last, we mention that the Gouy phase, which plays an important role in optics, also arises in the matter wave propagation [38, 39].

Finally, in this paper we have revisited the dissociation of H_2 molecules excited by electron impact, using a molecular nozzle beam in order to use molecules with cold internal degrees of freedom and high intensity flux [40]. Our setup adapts a high-precision time-of-flight spectroscopy experiment, with which we have studied the production of fast and slow metastable H(2s) atoms coming from the dissociation of cold H_2 molecules [41, 42]. This setup placed the detection system according to our collision kinematics analysis, positioning the detection active volume with very high precision to sample the direction where we believed that the twin atoms should be found.

DOUBLY EXCITED STATES OF THE HYDROGEN MOLECULE

Studies of the doubly excited dissociative channels of the hydrogen molecule have been performed since early 1960s. The first theoretical calculations of the doubly excited states were performed in 1974 [12]. In 2015, our group have make calculations of the potential energy curves for doubly excited states of the H_2 molecule in the 1.2a. u. - 12a. u. range at the multi-reference configuration interaction (MRCI) level, based on multiconfigurational self-consistent field orbital [19]. One of our goals, namely, to map the Q_2 state which dissociates in H(2s) + H(2s) was achieved successfully. Here, again, we should emphasize that no other technique was capable of providing this crucial information concerning applications in twin atoms experiments. In Fig. (**1**) several Q_2 potential curves are presented. Again, we should emphasize that no other technique was capable of providing this crucial information concerning applications in twin atoms experiments.

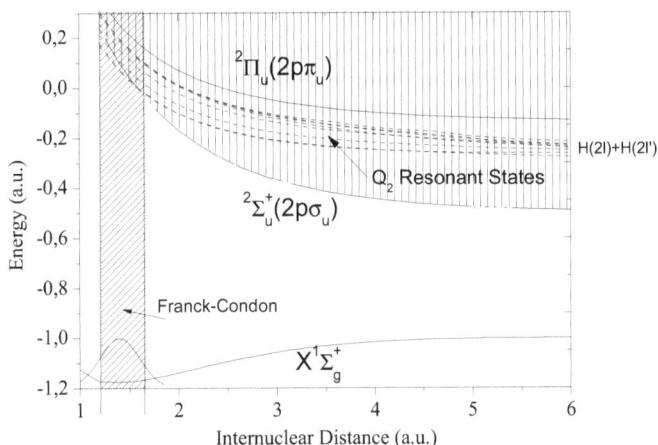

Fig. (1). Potential energy curves for doubly excited states of the H_2 - Q_2 potential curves.

Here we discuss our results for the potential curves of Q_2 states in the 1.2 a.u.-12.0 a.u. range. We have make generalized oscillator strength (GOS) calculations, within the vertical approximation. The 1.2 a.u. - 2.6 a.u. range for the H_2 molecule is the Franc-Condom region and is extremely important, especially in collisional processes, since this is the region where the electronic transitions take place. From the theoretical point of view, a reliable calculation of excitation cross sections requires that the wave functions are quite accurate in such a region.

EXPERIMENTAL SETUP

Here we briefly present one of our experimental setup (Fig. **2**) and then we discuss some results [40]. The apparatus consists of a supersonic jet source of molecular hydrogen, whose stream crosses an electron beam coming from a pulsed electron gun. The detection of the metastable atoms is performed by two detectors directly facing the collision zone, placed in opposite sides of the plane

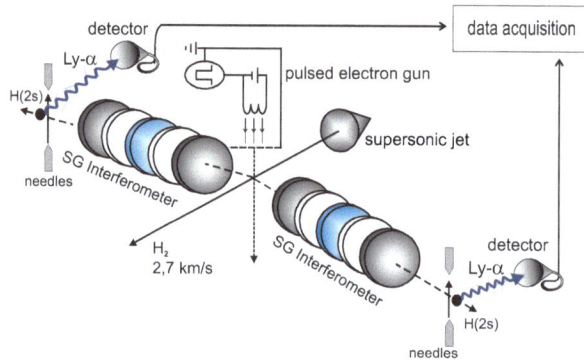

Fig. (2). Simplified sketch of the experimental setup.

defined by the electron and the H_2 beams. We have placed the detectors at different distances from the collision region, in order to avoid the coincident counts overlapping with any electric noise which could be picked up simultaneously by both detectors. In Refs [41, 42]. we were concerned with the precision of the time-of-flight spectra, which led us to use a specially designed detection system, which ensured we read only H(2s) atoms, and placed them as far as possible from the collision region. In the case of Ref [40]. our concerning was in enhancing the coincidence signals. Thus, we brought the detectors closer, placing them at 58 and 67 mm from the collision region and facing it directly. The effective area of the detectors leads to a solid angle of about 20 mile steroradian. Each detector (channeltron) was contained in a small Al box with a grounded grid shielding the entrance aperture.

The channeltron cone was biased negatively in order to avoid being saturated with scattered electrons. Since there is direct line of sight between the active area of the detectors and the collision region, we can detect, in addition to excited atoms, both UV radiation and ions, although we believe ions have very low probability of detection since there is little acceleration for them to gain energy and efficiently produce secondary electrons at the front surface of the detectors.

The time-of-flight (TOF) system enables us to separate the Lyman-α radiation emitted by $H(2^2S)$ from all the other contributions, since the Lyman-α coming from the metastable $H(2^2S)$ is produced in the detector's neighborhood where the quenching occurs, at larger times, whereas photons from $H(2^2P)$ will appear practically at the origin of the TOF spectra.

Now, let us discuss our coincidence measurement. The start of each sweep is established at the beginning of the electron beam pulse, while the signal coming from each detector feeds a different stop input into our time-of-flight acquisition card. Discrimination against noise is performed within the acquisition card by setting a "lower threshold" *via* software. Given a specific start signal, the arrival times provided by the stop's inputs of valid pulses are recorded with 16 ns resolution as integers corresponding to their bin number, where each detector furnishes a different column in the file. The analyzer card records all events where at least one stop for a given start was valid, and in this case a new line is added to the data file.

EXPERIMENTAL RESULTS

In Fig. (**3**) we display the top view of the coincidence counting spectrum (Fig. **3c**) with its corresponding TOF spectra placed accordingly to the axis associated with each one; on the bottom panel (Fig. **3a**) and on the left one (Fig. **3b**) we have the TOF spectrum registered by both detectors, A and B, respectively. On the very left of both graphs we can observe a vertical line which corresponds to the $H(2^2P)$ atoms, whose lifetime is of the order of 1 ns, and to the photons produced by the excitation of the radiative states of the hydrogen molecule. We observe that its width is of the order of 100 ns (the electron pulse width) as expected. In both figures (Figs. **3a** and **3b**) we see the (large) peak corresponding to the fast $H(2^2S)$ atoms.

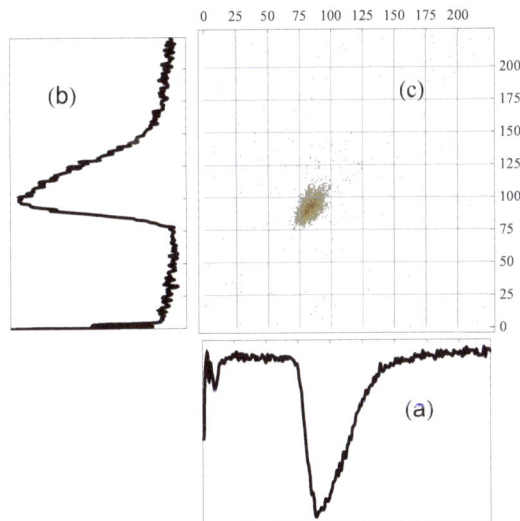

Fig. (3). Coincidence counting spectrum. Time-of-flight spectrum obtained with detectors A and B placed opposite of each other with respect to the collision plane and at distance of (a) 58 mm and (b) 67 mm from the H_2 dissociation spot. The H_2 excitation is accomplished by electron impact. The energy of the electrons is 200 eV and the electron pulse width is 100 ns. Time is displayed in units (bins) of 16 ns.

The peak of the slow atoms produced in the H_2 dissociation does not appear and it is further to the higher values of the bin scale. Fig. (**3**) shows the results of coincidence when the energy of the electrons is 200 eV and the electron pulse width is 100 ns. The horizontal and the vertical axis display time in units of 16 ns (bin) and this result was obtained during a round of the experiment which lasted 2 h and 8 min and yielded a total of 3164 coincidence counts.

The coincidence spectrum reveals random coincidences occurring over the whole domain in the bin plane with a very low count. It also shows a concentration of coincidences in a region which corresponds to the first third part of the fast $H(2^2S)$ peak on both detectors. In fact, the coincidences occurring in this region have a counting rate much higher than the average.

PROPOSAL

The experimental results and the theoretical calculations discussed in the present paper are good arguments to proceed in the direction pointed in the introduction. In fact, in reference [43] we proposed a scheme well suited to investigate quantitatively the angular-momentum coherence of molecular fragments. It is essentially a symmetric double Stern-Gerlach atom interferometer, which is suitable to estimate quantitatively the angular-momentum coherence of the fragments issued from a molecular dissociation. In spite of the several Stern-Gerlach atom interferometers accomplished to date, to our knowledge, no configuration able to treat this basic issue, concerning the spin coherence between two atoms, has been done neither proposed. We show that the envisioned double interferometer enables one to distinguish unambiguously a spin-coherent from a spin-incoherent dissociation, as well as to estimate the purity of the angular momentum density matrix associated with the fragments. This setup, which may be seen as an atomic analogue of a twin-photon interferometer, can be used to investigate the suitability of molecule dissociation processes —such as the metastable hydrogen atoms H(2s)-H(2s) dissociation— for coherent twin-atom optics.

In the next few paragraphs, we discuss in general terms the preparation of the H_2 molecule ground state and then its process of excitation.

Based on the symmetry properties of the nuclear and electronic wave-functions of the H_2, one can show that the anti-symmetric nuclear spin singlet state (I = 0), para-hydrogen, is associated with symmetric even J rotational levels and that the symmetric nuclear spin (I = 1), ortho-hydrogen, which corresponds to a nuclear triplet state, is associated with anti-symmetric odd J rotational levels. Hence the lowest energy state is the J =0 rotational state, which occurs only for the para form of the hydrogen. The first synthesis of pure para-hydrogen was achieved in 1929 [44] and since then the technique to produce pure para-hydrogen is well known [45].

Thus we assume that our system is initially composed of a para-hydrogen molecule, which is originally in the ground state with total angular momentum F_T = 0. We now discuss how to excite the molecule, without transferring angular momentum, to molecular states belonging to the so-called Q2 branch.

The desired di-excited states are not accessible – by electric dipole selection rule – from absorption of a single photon. In addition, using the Born approximation for electrons of high velocity, Bethe [46, 47] showed that, in the limit of large impact parameter (small linear momentum transfer), electrons behave like photons when they collide with a target, as well as they provide higher-order multi-pole excitation under conditions of small impact parameter (high linear momentum transfer). To our concern the desired states must be reached when the molecule is excited by an electron in a frontal collision, because it corresponds to a higher-order multipole transition together with a zero angular momentum transfer (s-wave limit). This condition is achieved by placing the detectors in the proper positions where only the fragments coming from frontal collisions can reach. These positions can be estimated by momentum and energy conservation. A detailed discussion of the collision kinematics regarding this type of experiment is given in ref [42]. Therefore, we assume that two atomic fragments, with a quantum number for the norm of the angular momentum equal to one, come out from the dissociation of an excited molecule with a null total angular momentum.

Thereby, in usual experimental conditions, the atomic fragments leave the collision region with almost opposite velocities of equal value much larger than the initial molecule CM velocity. As a direct consequence of our assumption of null initial total angular momentum, the atomic fragments leave the collision zone carrying away an opposite magnetic moment. Thus, one should design the double Stern-Gerlach interferometer as to select opposite projections of the transverse/longitudinal angular momentum on both sides.

Thus, after the fragmentation of the H_2 excited molecule, we have a system (the molecule in its asymptotic state) with a total angular momentum equal zero, F_T = 0, and consequently with its projection along any direction M_{FT} = 0, in particular along the molecular axis. Hence, let us consider the molecular axis in the center of mass frame, where the two atoms will have opposite velocities. In a simplified approach of the molecular state in its asymptotic region, we write the total wave function of a pure state as a product of the angular momentum and the spatial parts:

$$\left|\Psi_0\right\rangle = \left|\Psi(\vec{x}_1,\vec{x}_2)\right\rangle\left|F_T = 0, M_{F_T} = 0\right\rangle,$$

The spatial part is written in terms of the center of mass and relative motion of the two atoms. The center of mass motion is treated as a plane wave and the relative motion is described as a continuous vibrational wave function with zero angular momentum [29]:

$$|\Psi(\vec{x}_1,\vec{x}_2)\rangle = Ae^{i\vec{R}\cdot R}\frac{sin(kr+\delta)}{kr},$$

where $\vec{K}$ is the linear momentum of the center of mass, k is the modulus of the relative momentum related with the energy released E in the dissociation of the molecule, and δ is a phase factor which contains all the information of the repulsive potential curve of the di-excited molecule before its dissociation and the energy released E.

Next we consider that the detectors are placed in a macroscopic distance, so that we can consider that $kr \approx \vec{k}\cdot\vec{r} = \vec{k}\cdot(\vec{x}_1-\vec{x}_2)$. Thus, $|\Psi(\vec{x}_1,\vec{x}_2)\rangle$ yields

$$|\Psi(\vec{x}_1,\vec{x}_2)\rangle = Ae^{i\frac{K}{(m_1+m_2)}\cdot(m_1\vec{x}_1+m_2\vec{x}_2)}\frac{sin(\vec{k}\cdot\vec{r}+\delta)}{\vec{k}\cdot\vec{r}},$$

Using that $m_1 = m_2$, after a little algebra we get

$$|\Psi(\vec{x}_1,\vec{x}_2)\rangle = \frac{A}{kr}sin(\delta)e^{i\vec{k}_l\cdot\vec{x}_l}e^{i\vec{k}_r\cdot\vec{x}_r} = \frac{A}{kr}sin(\delta)\psi(\vec{x}_l)\psi(\vec{x}_r),$$

where we made the identification $\vec{k}_l = \frac{1}{2}\vec{K}-\vec{k}$ and $\vec{k}_r = \frac{1}{2}\vec{K}+\vec{k}$ as the momentum of the particle which goes to left and to right, respectively. Therefore we label the particles as *left* and *right*, and no more as 1 and 2: $\vec{x}_l$ and $\vec{x}_r$.

We suppose that each atom passes through a Stern-Gerlach Interferometer, (see Fig. **4**). The interferometers lie in opposite directions in the H_2 center of mass frame and constitute the two branches of our experimental setup. At first, along the interferometer, each atoms crosses a Lamb-Retherford polarizer, where a magnetic field of 600G is applied, so that only atoms in the hyperfine structure states f = 1, m_f = 0, 1 of $2s_{1/2}$ remain [48]. An important point of our proposal is that the magnetic field directions in the two polarizers should be reversed. The reason of that is the following. First, the atomic angular momentum component m_{fi} is defined according to magnetic field direction experienced by the atom; at the same time, we are dealing with a system, whose global angular momentum projection (along any direction) is zero, *i.e.* $M_{FT} = 0$. Therefore, if we had an ideal polarizer on each branch and put both magnetic field in the same direction, we would be able to measure in coincidence only the $m_{fi} = 0$ atoms, whereas the components $m_{fi} = 1$ of both sides would be eliminated in the coincidence measurements. On the other hand, with the opposite quantization directions on each side, the global condition $M_{FT} = 0$ (for the molecular state in its asymptotic limit) will be satisfied by the two atoms with $m_{f1} = m_{f2} = 0$ as well as by the ones with $m_{f1} = m_{f2} = 1$, opening the possibility to verify the entanglement of the two hydrogen atoms.

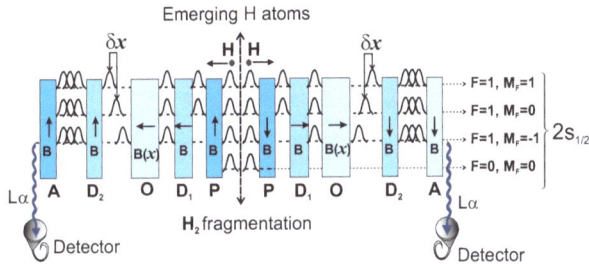

Fig. (4). The figure shows the basic elements of our proposed experimental setup (in the center of mass frame): the atomic polarizers **P**, which filter the hyperfine structure states $(f = 1, m_f = -1)$ and $(f = 0, m_f = 0)$; the beam dividers **D**, which split each of the remaining $(f = 1, m_f = 1)$ and $(f = 1, m_f = 0)$ states into a linear combination of the states $(f = 1, m_f = 1)$, $(f = 1, m_f = 0)$ and $(f = 1, m_f = -1)$; the phase objects **O**, which induce phase shift in each of the hyperfine structure states; the analyzers **A**, which mix the hyperfine structure state $(f = 1, m_f = -1)$ with the state $|2p_{1/2}, ñ$ in order to cause a decay by a Lyman-α, so that it can be measured. The arrows inside the boxes indicate the magnetic field direction in each device.

Hence, for a spin-coherent molecule dissociation, the quantum state of the system immediately after the dissociation would be $\left| F_T = 0, M_{F_T} = 0 \right\rangle$, (where F_T and M_{FT} are the quantum numbers for the total angular momentum of the two H atoms), or according to a Clebsh-Gordan decomposition [49] as

$$\left| F_T = 0, M_{F_T} = 0 \right\rangle = -\sqrt{\frac{1}{3}}\left|0_l 0_r\right\rangle + \sqrt{\frac{2}{3}}\left|1_l 1_r\right\rangle$$ up to a global phase. In this conditions and in

the center of mass frame, the total wave function of a pure state is given by

$$|\Psi_0\rangle = \frac{A}{kr}\sin(\delta)\,\psi(\vec{x}_l)\,\psi(\vec{x}_r)\left(-\sqrt{\frac{1}{3}}\left|0_l 0_r\right\rangle + \sqrt{\frac{2}{3}}\left|1_l 1_r\right\rangle\right)$$

$$= B(r,k,\delta)\left(-\sqrt{\frac{1}{3}}\psi(\vec{x}_l)\left|0_l\right\rangle\psi(\vec{x}_r)\left|0_r\right\rangle + \sqrt{\frac{2}{3}}\psi(\vec{x}_l)\left|1_l\right\rangle\psi(\vec{x}_r)\left|1_r\right\rangle\right)$$

In the expression above the coefficient $B(r, k, d)$ is a normalization factor which we will ignore it from now on:

$$|\Psi_0\rangle = \left(-\sqrt{\frac{1}{3}}\psi(\bar{x}_l)|0_l\rangle\psi(\bar{x}_r)|0_r\rangle + \sqrt{\frac{2}{3}}\psi(\bar{x}_l)|1_l\rangle\psi(\bar{x}_r)|1_r\rangle\right) = \left(-\sqrt{\frac{1}{3}}|\chi_0\rangle + \sqrt{\frac{2}{3}}|\chi_1\rangle\right)$$

The atomic fragments thus behave as an effective two-level system, whose coherence may be characterized by a Bloch vector. More generally, for a partially coherent dissociation, the angular-momentum degrees of freedom are described by an initial density operator of the form

$$\rho_0 = |\Psi_0\rangle\langle\Psi_0| = \frac{1}{3}|\chi_0\rangle\langle\chi_0| + \frac{2}{3}|\chi_1\rangle\langle\chi_1| - \lambda\sqrt{\frac{2}{3}}\left(|\chi_0\rangle\langle\chi_1| + |\chi_1\rangle\langle\chi_0|\right)$$

with $0 \leq \lambda \leq 1$. The parameter λ quantifies the spin coherence, indeed $\lambda = 1$ and $\lambda = 0$ correspond respectively to a completely spin-coherent and completely spin-incoherent dissociation. In what follows, we show the dependence of the interference pattern – obtained by coincidence detection after the double SG interferometer – with respect to the parameter λ.

Finally, we shall now discuss how our experimental sketch can distinguish the degree of coherence given by λ. The procedure involves a measurement of a physical quantity ($\hat{Q}$) such that the basis of states used, in our case the states $|1\rangle$ ($f = 1$, $m_f = 1$) and $|0\rangle$ ($f = 1$, $m_f = 0$), is not constituted by the eigenvectors of $\hat{Q}$. The corresponding quantity will be related to an operation consisted by the joint operations of the following devices: a Majorana region, also called beam divider, ($\hat{D}_1$); a phase object ($\hat{O}$); and another beam divider ($\hat{D}_2$), as shown in Fig. (**5**).

Thus $\hat{Q} = \hat{D}_2\hat{O}\hat{D}_1$ on each branch of the setup. The beam divider involves a non-adiabatic change of the quantization axis. It is basically a device in which the angular momentum quantization axis is abruptly changed, which causes each angular moment component to be given by a linear combination of all angular moment components belonging to the same vector subspace. We shall consider a rotation of $\beta_1 = \pi/2$. The phase object ($\hat{O}$) induces a phase shift on the wave function of each atom according to

$$e^{i\vec{k}_l \cdot \vec{x}_l} = e^{-ikx} \rightarrow e^{-ik\left(x - m_f \delta x_l\right)}$$

and

$$e^{i\vec{k}_r \cdot \vec{x}_r} = e^{+ikx} \rightarrow e^{+ik\left(x + m_f \delta x_r\right)},$$

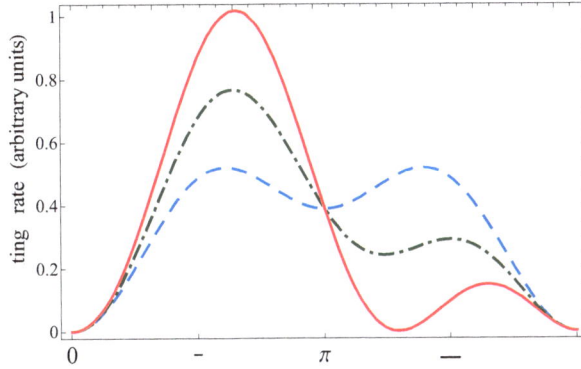

Fig. (5). (Color online) Counting rate corresponding to the simultaneous detection in the angular-momentum state $|-1_l, -1_r\rangle$. The blue curve (dashed line), green curve (dashed-dot line) and the red curve (full line) show the signal obtained for an initially incoherent ($\lambda = 0$), partially coherent ($\lambda = 1/2$) and fully coherent ($\lambda = 1$) density matrix, respectively. The left Zeeman phase shift has been fixed to $f_l = \pi/2$.

where $\vec{k}_l = -k\hat{x}$ and $\vec{k}_r = +k\hat{x}$ in the center of mass frame,

$$\delta x_{l(r)} = \left(g\mu_B \int B_{l(r)}(x)dx\right)/2E ,$$

with $B_{l(r)}(x)$ as the magnetic field of the phase object on the left (right) branch and E is kinetic energy of the atom. The second beam divider is such that the initial axis of quantization is restored. Formally, the detection signal can be analysed in terms of the projection of the density matrix onto the symmetrized quantum state $|-1_l, -1_r\rangle$; the counting rate is proportional to

$$I = \left\langle -1_l, -1_r \left| \hat{Q}_l \hat{Q}_r \rho_0 \left(\hat{Q}_l \hat{Q}_r\right)^\dagger \right| -1_l, -1_r \right\rangle,$$

whose expression is

$$I = sin^2 \frac{\phi_l}{2} sin^2 \frac{\phi_r}{2} \left[4\lambda \, sin \, \phi_l \, sin \, \phi_r + cos \, \phi_l \left(3 + 5cos \, \phi_r \right) + 3cos \, \phi_r + 5 \right],$$

where $f_l = kdx_l$ ($f_r = kdx_r$) are the atomic phases imprinted by the left (right)-side magnetic field on the atoms.

Fig. (**5**) shows the profile of the detection signal as a function of the phase f_r imprinted by the right phase object onto the atomic waves, for different values of the coherence parameter λ in the initial density matrix, and for a fixed value $f_l = \pi/2$ of the atomic phase provided by the left phase object. We can see that this experimental setup enables one to clearly distinguish a spin-coherent from a spin-incoherent fragmentation, and may allow to estimate the density matrix purity by measuring the asymmetry of the detection signal when the magnetic field of one of the phase objects is varied.

CONSENT FOR PUBLICATION

Not applicable.

CONFLICT OF INTEREST

The author (editor) declares no conflict of interest, financial or otherwise.

ACKNOWLEDGMENTS

This work is supported by CAPES, FAPERJ and CNPq.

REFERENCES

[1] G.H. Aguilar, A. Valdés-Hernández, L. Davidovich, S.P. Walborn, and P.H. Souto Ribeiro, "Experimental entanglement redistribution under decoherence channels", *Phys. Rev. Lett.,* vol. 113, no. 24, p. 240501, 2014.
[http://dx.doi.org/10.1103/PhysRevLett.113.240501] [PMID: 25541759]

[2] D. Bohm, *Quantum theory.* Prentice-Hall: Eglewood Cliffs, NJ, 1951.

[3] A. Einstein, B. Podolsky, and N. Rosen, *Phys. Rev.,* vol. 47, p. 777, 1935.
[http://dx.doi.org/10.1103/PhysRev.47.777]

[4] C.S. Wu, and I. Shaknov, *Phys. Rev.,* vol. 77, p. 136, 1950.
[http://dx.doi.org/10.1103/PhysRev.77.136]

[5] J.F. Clauser, and A. Shimony, *Rep. Prog. Phys.,* vol. 41, p. 1881, 1978.
[http://dx.doi.org/10.1088/0034-4885/41/12/002]

[6] A. Aspect, J. Dalibard, and G. Roger, *Phys. Rev. Lett.,* vol. 49, p. 1804, 1982.
[http://dx.doi.org/10.1103/PhysRevLett.49.1804]

[7] A. Aspect, P. Grangier, and G. Roger, *Phys. Rev. Lett.,* vol. 49, p. 91, 1982.
[http://dx.doi.org/10.1103/PhysRevLett.49.91]

[8] A. Lafosse, M. Lebech, J.C. Brenot, P.M. Guyon, L. Spielberger, O. Jagutzki, J.C. Houver, and D.

Dowek, *J. Phys. At. Mol. Opt. Phys.,* vol. 36, p. 4683, 2003.
[http://dx.doi.org/10.1088/0953-4075/36/23/007]

[9]   H.J. Wörner, S. Mollet, Ch. Jungen, and F. Merkt, *Phys. Rev. A,* vol. 75, p. 062511, 2007.
[http://dx.doi.org/10.1103/PhysRevA.75.062511]

[10]   E. Melero, "Garcia and al", *J. Phys. At. Mol. Opt. Phys.,* vol. 39, p. 205, 2006.*J. Chem. Phys.,* vol. 106, p. 7720, 1997.
[http://dx.doi.org/10.1063/1.473773] *J. Chem. Phys.,* vol. 110, p. 6702, 1999.
[http://dx.doi.org/10.1063/1.478576]

[11]   M. Leventhal, R.T. Robiscoe, and K.R. Lea, *Phys. Rev.,* vol. 158, p. 49, 1967.
[http://dx.doi.org/10.1103/PhysRev.158.49]

[12]   Y.V. Vanne, A. Saenz, A. Dalgarno, R.C. Forrey, P. Froelich, and S. Jonsell, *Phys. Rev. A,* vol. 73, p. 062706, 2006.
[http://dx.doi.org/10.1103/PhysRevA.73.062706]

[13]   F. Martín, J. Fernández, T. Havermeier, L. Foucar, T. Weber, K. Kreidi, M. Schöffler, L. Schmidt, T. Jahnke, O. Jagutzki, A. Czasch, E.P. Benis, T. Osipov, A.L. Landers, A. Belkacem, M.H. Prior, H. Schmidt-Böcking, C.L. Cocke, and R. Dörner, "Single photon-induced symmetry breaking of H2 dissociation", *Science,* vol. 315, no. 5812, pp. 629-633, 2007.
[http://dx.doi.org/10.1126/science.1136598] [PMID: 17272717]

[14]   R.C. Forrey, R. Côté, A. Dalgarno, S. Jonsell, A. Saenz, and P. Froelich, "Collisions between metastable hydrogen atoms at thermal energies", *Phys. Rev. Lett.,* vol. 85, no. 20, pp. 4245-4248, 2000.
[http://dx.doi.org/10.1103/PhysRevLett.85.4245] [PMID: 11060609]

[15]   D. Landhuis, L. Matos, and S.C. Moss, "J., Steinberger, K. Vant, L. Willmann, T. J. Greitak, and D. Kleppner", *Phys. Rev. A,* vol. 67, p. 022718, 2003.
[http://dx.doi.org/10.1103/PhysRevA.67.022718]

[16]   N. Kouchi, M. Ukai, and Y. Hatano, *J. Phys. At. Mol. Opt. Phys.,* vol. 30, p. 2319, 1997.
[http://dx.doi.org/10.1088/0953-4075/30/10/008]

[17]   J.L. Terry, *J. Vac. Sci. Technol. A,* vol. 1, p. 831, 1983.
[http://dx.doi.org/10.1116/1.572005] K.W. Gentle, *Rev. Mod. Phys.,* vol. 67, p. 809, 1995.
[http://dx.doi.org/10.1103/RevModPhys.67.809]

[18]   E. Schrödinger, *Proc. Camb. Philos. Soc.,* vol. 31, p. 555, 1935.
[http://dx.doi.org/10.1017/S0305004100013554]

[19]   L.O. Santos, A.B. Rocha, R.F. Nascimento, N.V. de Castro Faria, J. Ginette Jalbert, and B. Phys, "At", *Mol. Opt. Phys.,* vol. 48, p. 185104, 2015.
[http://dx.doi.org/10.1088/0953-4075/48/18/185104]

[20]   R. Lopes, A. Imanaliev, A. Aspect, M. Cheneau, D. Boiron, and C.I. Westbrook, "Atomic Hong-O--Mandel experiment", *Nature,* vol. 520, no. 7545, pp. 66-68, 2015.
[http://dx.doi.org/10.1038/nature14331] [PMID: 25832404]

[21]   C.K. Hong, Z.Y. Ou, and L. Mandel, "Measurement of subpicosecond time intervals between two photons by interference", *Phys. Rev. Lett.,* vol. 59, no. 18, pp. 2044-2046, 1987.
[http://dx.doi.org/10.1103/PhysRevLett.59.2044] [PMID: 10035403]

[22]   S. Michael, "Chapman, Christopher R. Ekstrom, Troy D. Hammond, Richard A. Rubenstein, JörgSchmiedmayer, Stefan Wehinger, and David E. Pritchard", *Phys. Rev. Lett.,* vol. 74, p. 4783, 1995.
[PMID: 10058598]

[23]   B. Brezger, L. Hackermüller, S. Uttenthaler, J. Petschinka, M. Arndt, and A. Zeilinger, "Matter-wave interferometer for large molecules", *Phys. Rev. Lett.,* vol. 88, no. 10, p. 100404, 2002.
[http://dx.doi.org/10.1103/PhysRevLett.88.100404] [PMID: 11909334]

[24]   A.E. Siegman, Lasers, (University Science Books, Mill Valey, 1990.

[25]   C.J. Bordé, *Fundamental Systems in Quantum Optics, Les Houches Lectures LIII*. Elsevier: New York, 1991.

[26]   C.J. Bordé, and C.R. Acad, *Sci. Paris,* vol. 4, p. 509, 2001.

[27]   A. Peters, K.Y. Chung, and S. Chu, *Nature,* vol. 400, p. 849, 1999. [http://dx.doi.org/10.1038/23655]

[28]   B. Canuel *et al.*, *Phys. Rev. Lett.,* vol. 97, p. 010402, 2006. [http://dx.doi.org/10.1103/PhysRevLett.97.010402] [PMID: 16907358]

[29]   F. Impens, P. Bouyer, and C.J. Bordé, *Appl. Phys. B,* vol. 84, p. 603, 2006. [http://dx.doi.org/10.1007/s00340-006-2399-3]

[30]   F. Impens, and C. J. Bordé, Phys. Rev. A vol.80, p.031602 (R), 2009

[31]   F. Impens, F.P. Dos Santos, and J. Christian, "Bordé", *New J. Phys.,* vol. 13, p. 065024, 2011. [http://dx.doi.org/10.1088/1367-2630/13/6/065024]

[32]   J-F. Riou, W. Guerin, Y. Le Coq, M. Fauquembergue, V. Josse, P. Bouyer, and A. Aspect, "Beam quality of a nonideal atom laser", *Phys. Rev. Lett.,* vol. 96, no. 7, p. 070404, 2006. [http://dx.doi.org/10.1103/PhysRevLett.96.070404] [PMID: 16606065]

[33]   F. Impens, *Phys. Rev. A,* vol. 77, p. 013619, 2008. [http://dx.doi.org/10.1103/PhysRevA.77.013619]

[34]   F. Impens, and Ch. Bordé, *Phys. Rev. A,* vol. 79, p. 043613, 2009. [http://dx.doi.org/10.1103/PhysRevA.79.043613]

[35]   J-F. Riou, Y. Le Coq, F. Impens, W. Guerin, C.J. Bordé, A. Aspect, and P. Bouyer, *Phys. Rev. A,* vol. 77, p. 033630, 2008. [http://dx.doi.org/10.1103/PhysRevA.77.033630]

[36]   F. Impens, *Phys. Rev. A,* vol. 80, p. 063617, 2009. [http://dx.doi.org/10.1103/PhysRevA.80.063617]

[37]   Ch.J. Bordé, *EPJD,* vol. 163, p. 315, 2008.

[38]   I.G. da Paz, M.C. Nemes, S. Padua, C.H. Monken, and J.G. Peixoto de Faria, *Phys. Lett. A,* vol. 374, p. 1660, 2010. [http://dx.doi.org/10.1016/j.physleta.2010.02.036]

[39]   I.G. da Paz, P.L. Saldanha, M.C. Nemes, and J.G. Peixoto de Faria, *New J. Phys.,* vol. 13, p. 125005, 2011. [http://dx.doi.org/10.1088/1367-2630/13/12/125005]

[40]   J. Robert, F. Zappa, C.R. de Carvalho, and R.F. Ginette Jalbert, "Nascimento, A. Trimeche, O. Dulieu, Aline Medina, Carla Carvalho, and N. V. de Castro Faria", *Phys. Rev. Lett.,* vol. 111, p. 183203, 2013. [http://dx.doi.org/10.1103/PhysRevLett.111.183203] [PMID: 24237516]

[41]   A. Medina, G. Rahmat, C.R. de Carvalho, G. Jalbert, F. Zappa, R.F. Nascimento, R. Cireasa, N. Vanhaecke, I.F. Schneider, N.V. de Castro Faria, and J. Robert, *J. Phys. B,* vol. 44, p. 215203, 2011. [http://dx.doi.org/10.1088/0953-4075/44/21/215203]

[42]   A. Medina, G. Rahmat, G. Jalbert, R. Cireasa, F. Zappa, C.R. de Carvalho, N.V. de Castro Faria, and J. Robert, *Eur. Phys. J. D,* vol. 66, p. 134, 2012. [http://dx.doi.org/10.1140/epjd/e2012-20657-8]

[43]   C.R. de Carvalho, and F. Ginette Jalbert, "Impens, J. Robert, J., Aline Medina, F. Zappa and N. V de Castro Faria", *Europhys. Lett.,* vol. 110, p. 50001, 2015. [http://dx.doi.org/10.1209/0295-5075/110/50001]

[44]   K.F. Bonhoeffer, and P. Harteck, *Z. Phys. Chem., B Chem. Elem. Proz. Aufbau Mater.,* vol. 4, p. 113,

1929.

[45] F.T. Wall, "Chemical Thermodynamics (W. H. Freeman and Com-pany, S. Francisco) 1974; L. A. Alekseeva and D. N. Kazakov", *Phys. Solid State,* vol. 49, p. 2104, 2007.

[46] H. Bethe, *Ann. Phys. (Leipzig),* vol. 397, p. 325, 1930.
      [http://dx.doi.org/10.1002/andp.19303970303]

[47] R.A. Buckingham, *Quantum Theory, I. Elements.,* D.R. Bates, Ed., Academic: New York, 1961.

[48] J. Robert, Ch. Miniatura, F. Perales, G. Vassilev, V. Bocvarski, J. Reinhardt, J. Baudon, and V. Lorent, *Europhys. Lett.,* vol. 9, p. 651, 1989.
      [http://dx.doi.org/10.1209/0295-5075/9/7/007]

[49] D.A. Varshalovich, and A.N. Moskalev, *Quantum Theory of Angular Momentum.* World Scientific Publishers: Singapore, 1988.
      [http://dx.doi.org/10.1142/0270]

SUBJECT INDEX

www.ingramcontent.com/pod-product-compliance
Lightning Source LLC
Chambersburg PA
CBHW050830220326
41598CB00006B/346